CW00547524

MOSLAH
MOSLAH
MOSLAH
MOSLAH
MOSLAH
MOSLAH
MOSLAH

SCOOTERS!

Michael Dregni & Eric Dregni
Foreword by Robert H. Ammon

Motorbooks International
Publishers & Wholesalers ®

First published in 1995 by Motorbooks International Publishers & Wholesalers, 729 Prospect Avenue, PO Box 1, Osceola, WI 54020 USA

Library of Congress Cataloging-in-Publication Data Available
ISBN 0-7603-0072-0 (pbk.)

On the front cover: The 1954 Piaggio calendar took *Vespisti* on a tour of the world, from Rome to Egypt to New York—Christmas shopping in Gotham was made easy by the Vespa 125cc.

On the frontispiece: Just the motorscooter for the dapper man about town: the Lambretta Li125 had class, styling, and a single-cylinder 125cc two-stroke engine. *Collezione Vittorio Tessera*

On the title page: Those magnificent men of the Moslah Temple Motor Corps and their scootering machines. Based out of Forth Worth, Texas, these fez-bedecked Shriners rode glorious white Cushman Eagles outfitted with most every Cushman Shriner option available. *Photo courtesy Shriner Sam Nelson*

On the back cover: Shriner Bill Ammon on a 1962 Cushman Super Eagle. *Photo courtesy Shriner Sam Nelson;* the Nibbio was the first Italian scooter to hit the market following World War II; this 1961 Lambretta Slimline Li125 was badge-engineered as a Lambretta Riverside Li125 and carried by Montgomery Ward department stores. *Owner: Eric Dregni*

Printed and bound in the United States of America

Contents

Acknowledgments

Being the scribes who put down on paper with quill and ink the history and mythology, as it were, of the motorscooter is a difficult and often thankless task. Difficult not so much because the threads in the web of history are lost in the mists of golden time, but more often because no one gives a hoot. Thankless because most people simply smirk when you so much as mention scoots.

Still we toil late into the night by the light of a solitary candle, striving to uncover the correct compression ratio of the first Cushman Auto-Glide or find out which Vespa model was Gina Lollabrigida's favorite. The task is made worthwhile because you, dear readers, are crazy for putt-putts just as we are.

No book writes itself, however, and no authors are solely responsible for the book that bears their names alone. We owe thanks to the following scooteristi and accessories to the crime, listed here in alphabetical order:

Kris Adams, the mod-est Mod; Robert H. Ammon; Bruno Baccari; Lindsay Brooke; Scott Chain of Scooterville USA; Sesostikis Temple Motor Patrol Captain Jack Douglass; François-Marie Dumas; Eric Dutra of Scootermania!; Giovanni and Anna Erba; Walt Fulton; Ray Gabbard; Tim Gartman and J.J. Gauthier; David Gaylin and Motor Cycle Days; Randolph Garner; Mustang maven Michael Gerald; Curt Giese; Ole Birger Gjevre; Noble Robert Gourlie; Gösta Karlsson; Jim Kilau, the man with the secret basement full of amazing scooters; Jeremy Leibig; W. Conway Link and his Deutsches Motorrad Registry; Noble Mark MacGillivray; Philip McCaleb; Roger McLaren; Vince Mross and West Coast Lambretta Works; Noble Sam Nelson; Sam Niskanen; Will Niskanen; Stefano Pasini; Sam Pitmon and the Pacesetters Scooter Society; Mark Preston; Imperial Shriner Photographer Tom Rousseau; E. Foster Salsbury; Herb Singe for loan of fantastic material from his collection of advertisements, brochures, and memorabilia from the early years of American scootering; Wallace Skyrman; Vittorio Tessera, keeper of the Lambretta flame; Keith and Kim Weeks for their incredible collection of American scooters; Noble Rick Watt; Noble Frank Workman; Steven Zasueta; and, of course, Zachary Zniewski.

And for keeping us running, our humble thanks to Dunn Bros. Coffee Shop for the best go-juice this side of the pond; Ted Cook's 19th Hole for the best Bar-B-Q north of Clarksdale, Mississippi, because man can not live on sushi alone; Hymie's Vintage Record City for 1950s blues on vinyl; and Willie's American Guitars because life's too short to live it with a crummy gee-tar.

Our deepest gratitude to the fearless publishing staff at Motorbooks International, especially Zack Miller, Jane Mausser and Amy Huberty, who managed to keep straight faces (at least in our presence) whenever the subject of scooters was raised.

Thanks, finally, to Sigrid Arnott and Nicolino.

◂Amore on Two Wheels
Love and Samson *motorrollers* went hand in hand in the 1950s. *W. Conway Link/Deutsches Motorrad Registry*

Foreword

The circumstances that led to building the prototype of the first Cushman motorscooter were interesting to say the least.

The aviator Colonel Roscoe Turner endorsed the Motor Glide scooter that E. Foster Salsbury had created in early 1936 in Los Angeles. Sometime in that same year, Roscoe Turner brought his air show to Lincoln, Nebraska, just down the road from the Cushman headquarters in Omaha. Turner toted this little Evinrude-powered Salsbury along with him. A neighborhood kid saw this scooter and decided that it would be fun to have one of his own. He found some angle iron and wheelbarrow wheels, and built himself a motorscooter. And he used a Cushman "Husky" engine from a lawn mower to power it.

We at Cushman found out about this scooter by chance. One day my dad, Charles Ammon—known affectionately to Cushman employees as "Uncle Charlie"—was at our spare parts depot in Lincoln and he looked out the window and saw the kid on his scooter buying parts. We were in the business of building and selling engines, and this scooter inspired my dad with the idea of making motorscooters to build and sell more engines. He charged me, with the help of our engineering department, to this task.

We built that first Cushman scooter when I was nineteen years old. The frame was made from 1 1/4in angle iron, but we soon learned that it wasn't strong enough. We then used 2in channel iron, which did the job. We learned early on that a rigid frame was essential, and that channel-iron design lasted a long time for Cushman. We used wheels from a wheelbarrow mounted with 4.00x8in tires, put an engine in it, and had it running in about thirty days.

I had read an article from the Engineering School at the University of Nebraska about the correct geometry for a bicycle or motorcycle. I developed the first scooter's steering head angle so that the tail end of the steering head was at a point behind where the wheel touches the ground. When I got the scooter stable enough so I could drive it with no hands, I knew it was ready.

Our goal was to build just a good, sturdy, two-wheel scooter. It was a very crude-looking thing, of course. But from those couple of prototypes came the first Cushman Auto-Glide, which went into production in late 1936. With the success of our first scooter, we continued to build motorscooters until Cushman was the largest scooter maker in the United States.

I first talked with Michael Dregni and Eric Dregni about five years ago when they were compiling information for their first book, *Illustrated MotorScooter Buyer's Guide*, which included a detailed history of Cushman and the Cushman motorscooters. Their first book also told the story of the development of American scooters and their influence on the scooters that were produced throughout the rest of the world following World War II, particularly the famous Vespa and Lambretta.

This second book tells more of that story in further detail and includes an amazing collection of around 400 photographs. I recommend it.

Robert H. Ammon
Paradise Valley, Arizona
March 1995

▸ **Shriner Bill Ammon**
Bill Ammon, brother of Bob Ammon and son of Cushman leader Charles Ammon, was a Shriner and instrumental in sparking the longstanding Cushman-Shriner relationship. Here he is in 1962 on a Cushman Super Eagle as Captain of the Sesostris Temple Motor Corps. *Photo courtesy Shriner Sam Nelson*

Introduction: Mondo Scooter

In the beginning, there was the Big Bang, the molten masses cooled and the galaxies formed, life emerged from the seas, humans developed, and in 1915, the first motorscooter was created. As of 1995, more than 20 million scooters have seen the light of day.

It began with a Big Bang. It continues with a putt-putt.

Throughout the ages, motorscooters have meant different things to different folk. Scoots have been cheap transportation for people without enough pennies for a car. They have been rebellious hot rods for misunderstood Wild Ones. They have carried lovers and warriors, priests and poets.

Even today, when motorscooters inspire collectors to go in search of their mythology and memorabilia, its different strokes for different putt-puttniks. To some, it's nostalgia for the classic Cushman "Turtlebacks" and Eagles and a remembrance of things past, those long-ago golden buzz-cut days of the American summer. To others, it's the full-bore two-stroke roar of a Vespa or Lambretta in anger, crowned by enough headlamps to light the night.

▲ **Scooters à la Grec**
Any time is ouzo time—as well as Vespa time—on the island of Naxos, Greece.

Scooters are funny. They are mechanical marvels on two wheels. Streamlined spuds. Mutant oddballs of Jet Age styling gone berserk. Innovative inventions shoehorned like sardines into miniaturized monocoque bodies. Engineering and styling enigmas, the stranger the better. They are two-wheeled pogo sticks, Italian hairdryers, Pushmans, dustbins on wheels, motorized lemons. Their names can be swear words, or their names can be uttered in worship by the faithful. They are the weird and the wonderful. They are the cute, the quaint, and the curious.

Explaining what inspires the scooter faithful ain't easy either. Tell a pal you like scoots and some hoot, some holler, most all smirk. Others nod and immediately launch into a dialectical discussion of the brake swept area on the first generation

► **Motorscooter Memorabilia**
Scootering's glory days from the IceCapades to bullfights to the Blind Lizard Picnic.

1' JAMBOREE EUROPEO LAMBRETTA '68
Lambretta club
paradiso per due
Vespa diventano
insuperabili nella G.S.
PARILLA
MV

▲ Ride, Ride, Ride...
Vespisti at Eurovespa 1962 in Madrid, Spain, ready themselves for a ride. *Scootermania!*

of 98cc Vespas. Explanation is not needed. If you have to ask, you just ain't gonna understand—and most Harley-Davidson folk don't have a clue what we're talking here.

Explaining what motorscooters have meant to society in the past 100 years is also something most people couldn't care a hounddog's flea about. To them it's all a tale told by an idiot full of sound and fury—although sound and worry may be closer to the truth.

Still, that truth is as weird and wonderful as a scooter itself.

> "Scooters were only a fashion even though it lasted for several years. Plenty of scooters had been made with little or no sales success until the Lambretta appeared and just why a device so ugly and awkward looking became a rage almost overnight, is something of a mystery."
> —Vincent motorcycle designer and scooter curmudgeon Philip Irving, *Black Smoke*, 1978

Product Liability Lawsuits On Wheels

Scooters have played a profound role in the world. Not only did they inspire the development of product liability laws, but they were also a missing link in the development of both riding lawn mowers and golf carts. And that's not all.

Scooters were born of bad times, economic depressions, and down-on-our-luck decades. The first scooter that truly made its mark was the Salsbury Motor Glide, created in the United States during the Great Depression by scooter visionary E. Foster Salsbury as cheap transportation for the depression-weary masses.

The second great scooter, Piaggio's Vespa, bowed onto the Italian scene in the days after the dust cleared from World War II. The Vespa had the same mission as Salsbury's Motor Glide, but the Vespa eventually traveled further. Alongside Innocenti's Lambretta and numerous other scooters, the Vespa and its kin provided the wheels beneath Europe and Japan's reconstruction.

Even today, scooters continue to provide cheap and cheerful transportation to much of the world, including India, the planet's largest scooterized society, still riding decades-old Lambrettas and Vespas and continuing to build new versions of the venerated elders.

On The Road To A Better World

Along the way and just by happenstance, scooters transformed society. For a window of just a few years following World War II, scooters played their part in reviving the European and Japanese economies before new cars and trucks

▲ **Putt-Putt Dreamer**
A genealogy of human beings and wheels, starting with a baby carriage. Graduating from the pram, the wee tot moves on to terrorize the town with a push scooter. Just add an engine, and the push scooter becomes a motorscooter, the next object of desire in many a child's dream. *Herb Singe Archives*

▲ **All For One and One For All!**
Three of the greatest motorscooters to ever ride Planet Earth. Scooters are a blend of styling, speed, and eccentricity. If they have to be explained, you just won't understand. These three are owned and were restored by scooter *impresario straordinario* Vittoro Tessera of Milan, Italy. From left, with drum roll and applause, please: styled like a two-tone primped poodle, the fabulous French-built Terrot; center, the scooter that saved the world, a 1950s 125cc Piaggio Vespa; right, the putt-putt to make Buck Rogers green with envy, America's Salsbury Super-Scooter Model 85.

were being mass produced. And along with hauling wares to market, scooters carried Romeo in search of Juliet, transported the whole family to the beach on the weekend, brought the city to the country and the country to the city. All along its route, the scooter helped disseminate ideas and culture far and wide.

Scooters also allowed women a newfound freedom of mobility, which, no matter how hard people tried, also gave women a newfound freedom of expression, exposures to new ideas, and a vision of broader horizons. We've come a long way, and scooters carried us part of the way.

Ironically, the motorscooter's history in the US of A was very different from that on the rest of the planet. Born in the USA, the motorscooter movement was some of the earliest of American-made popular culture to spread throughout the world, dating back to the first license-built version of the 1915 Motopeds built in Germany. But in the land of land yachts, scooters never truly caught on, always being considered putt-putt toys or hobbyist vehicles.

The Pope, The Who, And The IceCapades

Throughout history, it's easy to dismiss the scooter's role as being as small as the scooter's own elfin wheels, but motorscooters moved mountains one rock at a time.

Scooters have driven around the world. Broken land speed records. Sailed the high seas and crossed the English Channel fitted with pontoons and propeller. Danced the tarantella. Fought in bullfights. Been blessed by

▲ Saint Nick and Helper on a Vespa
The motorscooter helped modernize the world, providing inexpensive wheels to carry Americans during the Great Depression, Europe and Japan following World War II, and much of the Third World today. Even Santy Claus traded in his sleigh on a Vespa.

the Pope. Parachuted behind enemy lines. Fitted with bazookas. Packed with plastic explosives and used as terrorist bombs. Carried Gregory Peck and Audrey Hepburn on a Roman holiday. Fought alongside the Mods in the Brighton Beach wars against the Rockers and their brutish BSAs. Inspired Pete Townshend to hurl his Rickenbacker through his Marshall stack. Given Bo Diddley wheels to travel with his gee-tar. Run road races, endurance races, and off-road races. Brought the family on vacation to Norway. Wowed Fourth of July parade-goers across America with the Shriner Motor Corps' daredevil antics. Fought Communism. Fought Democracy. Carried the Good Word. Transported the rebel, the priest, and the Queen of England. Even starred in the IceCapades.

So without further ado, here's the humble tale of the motorscooter.

CHAPTER 1

The PIONEER MOTORSCOOTERS
1915-1930

It was the dawn of the Age of Speed, and everyone everywhere was in search of a vehicle to propel them forward at speed. The horse-drawn carriage—when fitted with a gas-, electric-, or even steam-powered engine—begat the automobile. The bicycle—when bolted to one of those cantankerous, oil-spewing "infernal"-combustion motors—begat the "motocycle," as it was termed by Indian in its early days. And the children's push scooter—when mounted with a smoke-coughing engine of some dubious repute to power either the front or the rear wheel in place of junior's hard-working leg—became the motorscooter.

Before the turn of the century, speed to most meant a horse or, to the lucky few, an iron horse. Who could imagine what kinds of speeds were even possible with the advent of Nikolas Otto's internal-combustion engine? The early auto pioneers—men like Frank Duryea, Henry Ford, Oscar Hedstrom—took horse-drawn carriages and bicycles and injected them with those new-fangled gasoline engines.

Everyone who was anyone had to have one of these powered vehicles, and the race was on. Ordinary citizens, who didn't truly know the meaning of the word "fast," suddenly couldn't get enough speed. In the coming years in the Age of Speed, speed would change the world, inspire architecture and industrial design in the form of Streamline Moderne that shaped everything from roadside pre-fab diners to coffeepots, and even create a new philosophy in the form of Futurism. Futurist Filippo Marinetti was infatuated with speed, as he wrote in his manifesto in 1909: "We say that the world's magnificence has been enriched by a new beauty: the beauty of speed. A racing car whose hood is adorned with great pipes, like serpents of explosive breath—a roaring car that seems to ride on grapeshot—is more beautiful than the Victory of Samothrace. . . . Time and Space died yesterday. We already live in the absolute, because we have created eternal, omnipresent speed."

The problem was that these first motorscooters never went very far. The early scooters were crude and clumsy, and often made it from point A to point B, but not back again. Alas, while development of the automobile and motorcycle sped ahead, the lowly scooter was left in the dust. Since these scooters were born from the child's scooter, it was an image the refined motorscooter was never able to outrun.

"Had the scooter lived, and not died stillborn as a result of makers rushing into production with untried and crude machines, we should probably have seen it develop into a most useful type of vehicle."
—*The Motor Cycle*, 1935

▸ **Cushman Glory Days**
From buzz-cut boyhood to fez-capped Shriner, the Cushman has symbolized scooters to America for more than fifty years.

Cushman MOTOR SCOOTER
Just as much fun as they look!
This is the year of the
SILVER EAGLE
with NEW high performance die cast aluminum engine
THE WORLD'S BEST BUY in low-cost transportation
DEALER INQUIRIES INVITED
LOW PURCHASE PRICE
LOW OPERATING COST
LOW MAINTENANCE
Cushman MOTOR WORKS, INC.
LINCOLN, NEBRASKA
THE IMPERIAL COUNCIL
ANCIENT ARABIC ORDER
OF THE
NOBLES OF THE MYSTIC SHRINE
FOR NORTH AMERICA
GENERAL OFFICES - TAMPA
CUSHMAN
Dedicated To The
Member
GREAT
MILE-
MAKING
MODELS

The First Scooter Craze

Mankind's Fall From Grace

The first of many subsequent scooter crazes burst onto the scene in the United States, England, and the European continent in the 1910s and 1920s. The American-made Motoped is believed to have been the first motorscooter to enter production, in the early 1910s; it was followed by the Autoped of 1914, created by the Autoped Company of New York. Meanwhile, in 1919, British engineer Granville Bradshaw created the ABC Scootamota; other pioneering English scooters included the Autoscoot and Autoglider in the 1920s.

The first motorscooters featured a small, raspy, oil-spewing engine mounted above the front or rear wheel, usually with chain drive to a wheel sprocket or direct roller drive running onto the tire. Some had gearboxes. Some had clutches. Many were single-speed and boasted of such weak power they did not even need a clutch. A tall steering rod ran from the front forks ending in handlebars often equipped with a throttle and brake lever. These early scooters had a central floorboard to stand on, usually just inches from earth, so the driver could put a foot down to the safety of solid ground should anything untoward happen.

"There was a sudden [scooter] boom. The idea of an inexpensive runabout for one caused a miniature furore, and a dozen or more firms got busy. Here, it *seemed,* was easy money."
—*The Motor Cycle,* 1935

◂ Motoped at Dizzying Speed
These brave souls actually went for a ride on their Motoped, although the look on the woman's face suggests less than overwhelming glee. With the bulk of the 1.5hp engine mounted on the left side of the front wheel, which was steered by that near-vertical spindle of a steering rod, more than one intrepid Motoped pilot must have visited terra firma gluteus maximus first. Fortunately, the fall from grace was not a long one, as the floorboard was only inches above ground so the driver could easily reach earth—a wise feature since the Motoped had no brakes. Riding such a scooter was not for the faint of heart.

▶ 1914 Autoped
"Wonder of the Motor Vehicle World" can only begin to describe the Autoped. The idea was simple: Borrow junior's push scooter, bolt a miniature engine to the front wheel with direct drive, fill the tank with gas, and you're ready to drive off into the future. The Autoped from the suitably named Autoped Company of New York and the similarly styled Motoped are believed to be the first motorscooters ever built, debuting in 1914-1915. One of the wonders of the Autoped and its scooter brethren must have been the ride. The Autoped was also license-built by Krupp in Germany with a 155cc four-stroke motor and by CAS in Czechoslovakia. Production of the Autoped halted in 1921.

THE AUTOPED

IS THE WONDER OF THE MOTOR VEHICLE WORLD

FIRST MODEL, CIRCA 1914

It's the only motor-driven passenger-carrying vehicle that can be carried into your office or home; so compact that it can be kept anywhere.

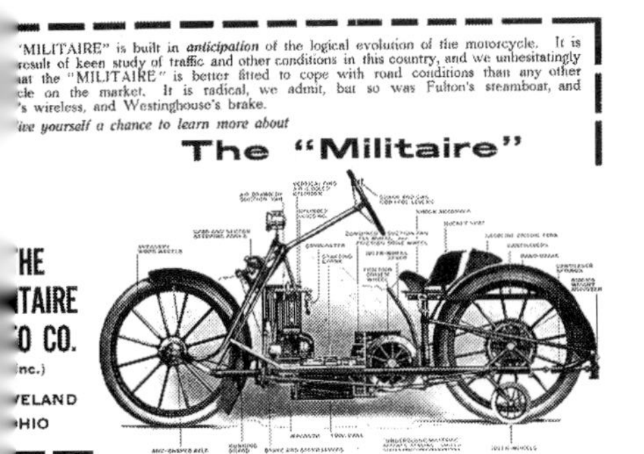

▲ 1912 Militaire
The American Militaire was part motorscooter, part motorcycle, and part Jules Verne spaceship, a far-out vision of the future in the pioneering days of motorized vehicles. The ad copy just begs to be quoted: "It is the result of keen study of traffic and other conditions in this country, and we unhesitatingly affirm that the 'Militaire' is better fitted to cope with road conditions than any other motorcycle on the market. It is radical, we admit, but so was Fulton's steamboat, and Marconi's wireless, and Westinghouse's brake." With these heady thoughts in mind, note the Militaire's artillery wood wheels, open-to-the-elements steering gears, shock absorber, bucket seat, rider's weight adjuster, and the idler wheels that could be lowered while waiting at America's first electric traffic signal, erected in Cleveland in 1914. Numerous other two-wheelers followed in the Militaire's image, including the American and English versions of the Neracar, the French Monet-Goyon Vélauto, German Megola and DKW Golem, and English Wilkinson. As this ad stated, "The 'Militaire' is built in *anticipation* of the logical evolution of the motorcycle"—a 1912 version of the 1950s sales pitch, "As new as tomorrow," which leads one to believe that even way back in the past, tomorrow was always here today—perhaps one day too soon.

▲ Amelia Earhart and Motoped
When she wasn't flying, pioneer aviator and feminist Amelia Earhart scooted about on her Motoped. Seeing Earhart with her scooter at the Burbank airport inspired visionary E. Foster Salsbury to create his revolutionary Motor Glide in the 1930s.

▲ Autoped and Doughboy
The US Army considered drafting the Autoped for patriotic service, and this brave doughboy tested the scooter in the mid-1910s. With this secret weapon to lead the bayonet charge out of the trenches, who knows what the outcome of World War I could have been—the Allies might even have lost! The exhaust fumes of a whole legion of scoots would probably have also gassed the good guys. Alas, the scooter would have to wait for World War II to see action. *Fred Crismon Archives*

A Putt-Putt For Pithy Ladies

Just the Thing to Ride to Tea

The first motorscooters were quaint, faddish vehicles at best, and they were not taken too seriously even at the time. Created for pithy ladies to putt-putt to town for shopping or afternoon tea, the scooter allowed Milady to stand upright so her neatly ironed dress did not suffer a wrinkle. Thus was born the motorscooter's first sales pitch.

The scooter was just a passing fancy in the Roaring Twenties, going the way of the Charleston and steam-powered automobiles. The first wave of scooters died by the side of the road as quickly as it was born.

▲ 1919 ABC Skootamota
This daring young lady on her ABC Skootamota fords a roaring river on her way to afternoon tea. Her trusty steed, the Skootamota, was at the forefront of a scooter revolution that swept Great Britain in 1919. Created in that same year by British engineer Granville Bradshaw, the Skootamota was powered by a 147cc intake-over-exhaust engine, later supplanted by a 110cc overhead-valve motor. The design of the Skootamota was ideal for ladies with its step-through frame making way for petticoats and frilly frocks. But the big advance heralded by the Skootamota over such machines as the Autoped and Motoped was the applauded addition of a seat. Couthness had come to the motorscooter, and pioneering machines such as the Skootamota set the style for scooters to come.

> "They vibrated to an extent that was almost unbelievable and their reliability was far from high, and they were not exactly comfortable with their small-diameter wheels and small-section tyres. The scooter was killed—by the scooter."
> —*The Motor Cycle*, 1935

▶ Tank-Treaded Roadless
Explain this, if you will. Yes, Victoria, it's a motorscooter with a tank tread. The British Roadless rode on a single narrow rubber track that allowed it to traverse most any terrain. Those handlebars turned the forward end of the track assembly, and the Douglas flat-twin with its 2.75hp powered the Roadless to a 20mph top speed. It was a scary thought.

▲ 1919 Autoglider De Luxe
Returning from the horrors of the First World War, many British soldiers were seized by a poetic wanderlust and took to the roads in search of the meaning of life. It was a plot that would be rehashed after World War II and lead to the birth of the Hell's Angels in California riding war-surplus Harley-Davidsons. In Great Britain, one result of this wanderlust was a scooter craze that lasted for a full summer, that of 1919, when everyone everywhere was talking of scooters. The Autoglider was crafted by Charles Ralph Townsend's Townsend Engineering Company of Birmingham. The first Autoglider Model A was standup style whereas this Model D featured a seat, which also provided the "suspension." The 269cc Villiers was mounted above, and drove the front wheel via a plate clutch. Such a scoot was an ideal wandering companion to ride to the end of the earth at Land's End and gaze at the sea and ask, "Why?" But could you ever envision Sonny Barger on it?

▲ 1920s Unibus and Bowler-Hatted Rider
"The Car on Two Wheels" concept was tried even way back in the 1920s! Built in England by the Gloucestershire Aircraft Company of Cheltenham, the Unibus disappeared as quickly as it arrived, debuting in 1920, going out of production by 1923. This may have disproved the Unibus' advertising slogan noted in this ad, "To see the Unibus is to want one." Unfortunately, not enough people did the latter. The world could have been a much better—or at least more interesting—place if they had.

CHAPTER 2

The GREAT AMERICAN SCOOTER CRAZE

1935-1940

The true ancestor of the modern motorscooter was assembled like Frankenstein's monster from odds and ends, bits and pieces, something borrowed, something blue. The creators were two young Californians, "financier" E. Foster Salsbury and inventor Austin Elmore. In this case, the mad creators' castle was the backroom of Salsbury's brother's heating and plumbing shop in Oakland. The year was 1935.

America was sunk in the midst of the Great Depression, but Salsbury had a vision of a cheap and cheerful vehicle that would propel the country forward to prosperous times. He witnessed the great aviator and feminist Amelia Earhart putt-putting about the Lockheed airport at Burbank on one of the rare Motopeds that was still alive and running. The vision was an inspiration: "It got me started thinking about building a real scooter," Salsbury remembered in 1992.

Salsbury's Motor Glide sparked a second scooter craze starting in the mid-1930s. The first craze that began with the Autoped and Motoped in 1915 had died away, and Salsbury's timing was perfect. Not only had twenty years passed and many people had forgiven—or at least, forgotten—the sins of past scooter builders, but now people actually had a need for what the motorscooter offered.

Where Salsbury led, others followed. After Salsbury spurned a bid for the Husky motors built by the Cushman firm in Lincoln, Nebraska, Cushman kickstarted its own Auto-Glide scooter into life in 1936—and created a scooter empire that eventually endured far beyond that of Salsbury. In Chicago, the Moto-Scoot and Mead Ranger scooters coughed to life in 1936; they were joined soon after by the Powell, Rock-Ola, Crocker Scootabout, Keen Power Cycle, the aptly named Puddlejumper, and many others.

This second generation of scooter makers took up where the scooter pioneers had left off—and paid heed to several important lessons that the pioneers had learned along the way. The internal-combustion engine was now much better developed, and these new scooters were more reliable machines with hardy engines. They boasted real gearboxes, brakes that actually braked, and after 1937, many had a radical torque converter transmission, again following Salsbury's lead. But most importantly for the scooter's future acceptance, the new scooters were dressed in bodywork that covered the engine and other dirty mechanical bits so drivers never had to soil their hands. The motorscooter had been civilized.

"The Salsbury Motor Glide is the greatest woman catcher I have ever seen."
—Air racer and barnstormer Colonel Roscoe Turner

▸ **Crazy for Scooters, Circa 1930s**
While innocent jazz rhythms wafted forth from the small-town park's gazebo, a revolution was in the making. Foster Salsbury's Motor Glide and Robert Ammon's Cushman Auto-Glide were born, and there was no escaping the future of the motorscooter. Guitar: 1933 National Style O with aluminum resonating disk in the days before electricity revolutionized the guitar world.

MOTOR GLIDE
PE'S ECLIPSING SENSATION
OUDINI
RLD'S HANDCUFF KING & PRISON BREAKER
"NOTHING ON EARTH CAN HOLD
1937
Aero Model
PERSONAL RUNABO
AGAVULIN
16
FINGERMAN
EXCLUSIVE PICTU
ETECTIVE
OLICE CASES
Glide
Over 120 Miles To The Gallon
More Than 30 Miles Per Hour
N MOTOR WORKS LINCOLN NEBR
MOONLIGHT LOVE — AND MUR

Salsbury Motor Glide

The Motorscooter That Started It All

E. Foster Salsbury had a vision. Inspired by the sight of aviator Amelia Earhart dashing around the Lockheed airport at Burbank, California, on an ancient Motoped, he decided to create his own scooter for the masses. Salsbury enlisted the aid of inventor Austin Elmore, and together the dynamic duo created the first Motor Glide.

The Motor Glide made its debut at the February 1936 Airplane and Boating Show in Los Angeles' Pan-Pacific Auditorium. Salsbury showed off his scooter to barnstorming aviator Colonel Roscoe Turner who was thrilled by the Motor Glide and became a Salsbury spokesman, helping the fledgling firm find a market.

▲ 1937 Salsbury Aero Model Motor Glide
Salsbury did not rest on its laurels. The fledgling firm wasted no time in updating its scooters, and in late 1936, the arrival of the new 1937 model was announced—less than a year after the Motor Glide's debut! It was now powered by a new Johnson four-stroke engine with a 2 1/8x1 3/4in bore and stroke for 6.2ci creating 0.75hp at 2400rpm. The Motor Glide operating instruction had only two steps: "First—Give the Motor Glide a push and step on. (Automatic compression release makes starting easy.) Second—Give it gas by opening the throttle with your right hand and away you go." Owner: Herb Singe.

◄ 1937 Salsbury Aero Model Motor Glide
The features of the Motor Glide were legion and legend, as pointed out by this later 1936 newspaper ad. "Dash hither and yon in gay abandon...glide out to the country club or to the beach," tantalized the superlative-loving ad writer. The Motor Glide parks on a dime, turns in a 3ft circle, "eases through traffic like an eel," and gets 150 miles to a gallon of gas. In addition, "it's as safe as an armchair and as comfortable...dependable as time; as easy-handling as a docile mare." You too could be gliding along instead of merely "puddlejumping." *E. Foster Salsbury Archives*

▸ 1936 Salsbury Motor Glide Number 1
A portrait of motorscooter firsts: Here is the first Salsbury advertisement, printed in magazines in February 1936, showing the first Salsbury Motor Glide with the company's first endorser, barnstorming aviator Colonel Roscoe Turner. The Colonel was touted in Salsbury ads as "America's best-known flying ace"; he was actually more of a showman and an honorary National Guardsman, but he got the point across. This first model differed from all subsequent Motor Glides in that the drive from the engine was via direct roller friction onto the rear tire. Power came from an Evinrude Speedibyke single-cylinder two-cycle engine with a 2x1 5/8in bore and stroke for 5.1ci. Churning along at 3500rpm, the engine created 0.75hp. As a note on the back of this ad from the Salsbury archives stated, concerning the roller drive, "It worked fine on dry pavement." Not exactly a golden endorsement from the company itself, but in a few months, Salsbury reversed rotation of its engine and chain drive was added as a remedy. The Motor Glide was now able to venture forth after a rain. Foster Salsbury estimated that only 25-30 of the roller-drive Motor Glides were ever built. *E. Foster Salsbury Archives*

Ride a MOTOR GLIDE!

COLONEL ROSCOE TURNER arriving home on his new Motor Glide after completing one of his recent airplane trips.

THE NEW MOTOR GLIDE was designed and built for Colonel Roscoe Turner to carry in his airplane for transportation to and from the many airports on which he lands.

Colonel Turner's MOTOR GLIDE has been so enthusiastically received by the Aviation World that it has become necessary to go into production on this model to supply the demand for this new sensational mode of transportation.

The public is beginning to realize that the MOTOR GLIDE is practical for many purposes in addition to that for which it is used by Colonel Turner. Ride a MOTOR GLIDE for a new Thrill! Glide along effortlessly mile after mile. Ride it to work; visit friends and interesting places. Use it for deliveries, collections and other business purposes. It costs so little. Ride the MOTOR GLIDE five miles for one cent. The lowest cost transportation available.

The MOTOR GLIDE is sturdy, safe, economical, comfortable and dependable. Built with the best standard parts and materials obtainable. Weighs 65 pounds and fully equipped with lights, horn, airwheels, etc. The handle bars are built to fold down so that it can be carried in your car, plane or boat.

Place your order now for early delivery. For further information call or write

THE MOTOR GLIDE COMPANY

OAKLAND—1450 Harrison St., HIgate 2480 • LOS ANGELES—1041 N. Sycamore Ave., HEmpstead 8131

February 1936

▴ 1937 Salsbury Aero Model Motor Glide Showroom
Salsbury's new Motor Glide was the state-of-the-art scooter in 1937, and this showroom display at Beebe Company in Portland, Oregon, featured three of the latest models. While early scoots had battery-powered bicycle headlamps bolted to the handlebars, Salsburys now boasted a full array of electrical gadgets, including a horn, headlamp, and taillamp—all powered by no less than six herculean 6-volt batteries. And while suspension was but a dream, the Motor Glide did have a seat that was "amply padded with felt and hair," making the scooter "extremely comfortable," according to Salsbury's ads. In fact, the Motor Glide was so cool, even Bing Crosby had one. *E. Foster Salsbury Archives*

"I had no idea what the market for a scooter would be. It was pure invention—very far out for those days."
—E. Foster Salsbury, 1992 Interview

Cushman Auto-Glide

Birth of a Motorscooter Empire

▲ 1937 Cushman Auto-Glide Model 1
Despite the halo surrounding the Auto-Glide, the motorscooter of 1937 was still a crude affair. Cushman's frame was welded up of 2x1/8in channel steel that amounted to a heavy-duty, heavy-weighing affair. "Suspension"—not a word that appeared in many scooter ads of the day—resulted from the 3.50x12in balloon tires, the padded seat, and the rider's gluteus maximus. So while engineering may not have been Jet Age, operation was easy: the "smooth-as-velvet" clutch to the single-speed drive was operated by the left foot, the rear-wheel drum brake was controlled by the right foot, and the handlebar-mounted throttle by the right hand. For optional accessories, a bicycle light could be bolted onto the handlebars.

▲ 1937 Cushman Auto-Glide Model 1
One of the earliest Cushman brochures for the firm's first production model, the aptly named Model 1, which made its soul-stirring debut in 1937. At the heart of the Auto-Glide was the Cushman Husky engine with a whopping 1.5hp worth of motive power. Bore and stroke were 2 5/8x2 1/2in for 13.53ci. Although it was the "Newest Riding Thrill," all that dizzying power added up to only a 30mph top speed, but Cushman, ever the clever marketeers, promoted the slow speeds in its ads: "Low speed ensures safety. Auto-Glide is not a 'speed wagon.'" Yes, but did you ever see Bing Crosby on one?

► 1939 Cushman Auto-Glide Model 2
"All aboard for a wonderful Glide Ride!" shouted this 1939 ad from the back pages of *Popular Mechanics*. The superlatives were flowing fast and furious to describe the virtues of the new Model 2 with its 2hp Husky engine and optional two-speed transmission. Cushman also introduced a new three-wheeled scooter, the Model 9 Package-Kar, which was a Model 2 with a two-wheeled axle and delivery box bolted on front. As the ad promised, the Auto-Glide was "ideal for work, school, deliveries—any place you'd use a car!"

The Cushman Motor Works of Lincoln, Nebraska, followed Salsbury's lead into the novel motorscooter market. In mid-1936, Salsbury requested—and then rejected—a bid for 1,000 of Cushman's famous Husky engines that he planned to use in his scooter. Cushman chief "Uncle Charlie" Ammon saw the writing on the wall and assigned his son, Robert, to create what became the Cushman Auto-Glide.

With the success of its first scooter, Cushman continued to construct the most long-lived American motorscooter in a dazzling variety of unique models. And as Vespa became synonymous for scooter in the rest of the world, Cushman stands for scooter to most Americans even today.

"We were in the business of building and selling engines. The idea of making a motorscooter was to build and sell more engines."
—Cushman chief Robert H. Ammon, 1992 Interview

◄ Ward Tractor Plow with Cushman Engine
Cushman's roots stretched back to 1901 when Everett and Clinton Cushman began building engines in their basement. Throughout the 1910s and 1920s, Cushman engines powered everything from racing boats to the Ward Tractor Plow. In 1922, Cushman created its famed Husky engine that powered the firm's Bob-A-Lawn mower. E. Foster Salsbury approached Cushman in 1936 requesting a bid for 1,000 Husky engines to power his Motor Glide, and then bought elsewhere. With a copy of Salsbury's blueprints in hand and inspired by a neighborhood kid's homebuilt scooter, Cushman welded up a prototype for its Auto-Glide motorscooter. Why let a good idea go to waste?

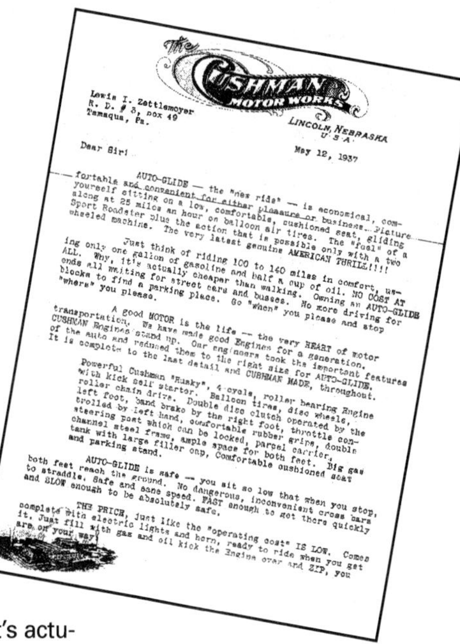
The Cushman Motor Works
Lincoln, Nebraska U.S.A.

Lewis I. Zettlemoyer
R. D. # 3, Box 49
Tamaqua, Pa.

May 12, 1937

Dear Sir:

AUTO-GLIDE — the "new ride" — is economical, comfortable and convenient for either pleasure or business. Picture yourself sitting on a low, comfortable, cushioned seat, gliding along at 25 miles an hour on balloon air tires. The "feel" of a Sport Roadster plus the action that is possible only with a two wheeled machine. The very latest genuine AMERICAN THRILL!!!!

Just think of riding 100 to 140 miles in comfort, using only one gallon of gasoline and half a cup of oil. NO COST AT ALL. Why, it's actually cheaper than walking. Owning an AUTO-GLIDE ends all waiting for street cars and busses. No more driving for blocks to find a parking place. Go "when" you please and stop "where" you please.

A good MOTOR is the life — the very HEART of motor transportation. We have made good Engines for a generation. CUSHMAN Engines stand up. Our engineers took the important features of the auto and reduced them to the right size for AUTO-GLIDE. It is complete to the last detail and CUSHMAN MADE, throughout.

Powerful Cushman "Husky", 4-cycle, roller bearing Engine with kick self starter. Balloon tires, disc wheels, roller chain drive. Double disc clutch operated by the left foot, band brake by the right foot, throttle controlled by left hand, comfortable rubber grips, double steering post which can be locked, parcel carrier, channel steel frame, ample space for both feet. Big gas tank with large filler cap, Comfortable cushioned seat and parking stand.

AUTO-GLIDE is safe — you sit so low that when you stop, both feet reach the ground. No dangerous, inconvenient cross bars to straddle. Safe and sane speed. FAST enough to get there quickly and SLOW enough to be absolutely safe.

THE PRICE, just like the "operating cost" IS LOW. Comes complete with electric lights and horn, ready to ride when you get it. Just fill with gas and oil kick the Engine over and ZIP, you are on your way!

► 1937 Cushman Letter
Dated May 12, 1937, this early promotional letter promised Mr. Lewis I. Zettlemoyer of Tamaqua, PA, "the very latest genuine AMERICAN THRILL!" Cushman promised safety and economy as its key sales points throughout its motorscooter ad campaigning, and this early letter set the stage. On the economy front, the Auto-Glide could go 100-140 miles ("in comfort"—phew!) on 1 gallon of gas and 1/2 cup of oil (that's an oil-consumption rate at which you would quickly rebuild most modern car engines!). Cushman noted, in original all-cap letters, that this was "NO COST AT ALL. Why, it's actually cheaper than walking," which was probably not too truthful a statement. Safety came from the slow "sane speed" of the Auto-Glide: "FAST enough to get there quickly and SLOW enough to be absolutely safe." *Herb Singe Archives*

Join the thousands who now go places for less!

Auto-Glide
CUSHMAN-Powered

Now, all aboard for a wonderful Glide Ride!—the most economical you ever enjoyed! Thousands from coast to coast get 125 MILES PER GALLON, 35 miles per hour. Ideal for work, school, deliveries—**any place** you'd use a car! Husky 1½ or 2 H.P. engine has fast get-away, zips up hills. Powerful brakes stop machine on a dime. Handles like car in traffic. Easy to park. Extra rugged construction. Safe on all roads. Clutch (most models) makes starts, stops, easy. 2-speed transmission (optional). Big, comfortable cushion, room for two!

5 NEW MODELS! Standard, (above) newest low-priced 2-wheel model for business, sport. De Luxe model, streamlined, electric lights, for youngsters, office workers. **KARI-PAC**, big rear carrier for deliveries, school, etc. **SIDE-KAR**, attaches to above models, makes 3-wheel unit. 3-WHEEL DELIVERY MODEL (left) has 400 lb. capacity.

125 Mi. PER GAL.

35 Mi. PER HR.

$92.45 UP

WRITE! Or see your dealer for Free book showing all models. Send NOW!

THE CUSHMAN MOTOR WORKS
Dept. T-3, Lincoln, Nebr.

Dealers: Get Spl. Offer Now!

Moto-Scoot and Mead Ranger

The Motorscooter Craze Moves East

◄ 1938 Moto-Scoot
Styling came to the Moto-Scoot in 1938 with a metal wicker-pattern engine cover that was years ahead of the other scooters on the market—in fact no one *ever* followed this look. Motive power from the new Lauson engine was a whopping 1.5hp via direct drive to the rear wheel without any of the tiresome hassles of a clutch. Suspension in the rear was by way of the bicycle seat's springs, probably pilfered from the Mead Ranger bicycle assembly line; up front, suspension was purely by the inherent flex and compression of solid metal. The long, thin cylindrical metal tube on the steering post held a stack of batteries that feebly powered the headlamp and taillamp. Braking was at the rear only and controlled by the bicycle-style handgrip lever—no one trusted front braking very much in those days, and probably rightly so as the cheap lining materials were prone to lock up if the brakes were applied too hard. Moto-Scoot top speed was now a breathtaking 30mph. Owner: Herb Singe

◄ 1937 Moto-Scoot
Norman A. Siegal had a golden vision of the future, a vision that rode on two diminutive wheels and was powered by an elfin engine. In 1936, the 27 year old hired three workmen, withdrew his banked savings, and rolled up his sleeves to begin building his Moto-Scoot. Siegal had obviously studied Salsbury's Motor Glide and done his homework—right down to naming his creation. As this June 1937 *Popular Mechanics* ad boasted, there was "No other like it!" but the only thing preventing a patent infringement lawsuit from Salsbury was that Salsbury didn't think about patenting his creation back then. If you didn't look closely, you might have missed the engine and taken the Moto-Scoot for a child's push scooter, but it is there: a 1/2hp Lauson mounted on the right side of the rear wheel with a shaft transferring chain drive to the left side.

"Moto-Skoot [sic] inventor Norman Siegal is the Henry Ford of the scooter business."
—*Time* magazine,
April 3, 1939

At about the same time as Cushman rolled out its Auto-Glide, the Moto-Scoot Company in Chicago was launching its new motorscooter as well. The prototype Moto-Scoot scooter was created by visionary Chicagoan Norman Siegal.

Nearby, the venerable Mead Company had been building bicycles since 1889, and had created a vast marketing network for its famous Ranger pedal cycle. By 1936, Mead was ready to follow the trend to motorized two-wheelers, and signed on the dotted line to have Siegal build badge-engineered Moto-Scoots as the new and novel Mead Ranger scooter. With its established sales avenues, the Mead became one of the best-selling early scooters and the Mead–Moto-Scoot deal continued through the 1950s.

=RUSH ORDER BLANK=
MEAD CYCLE CO., 17 S. Market St., Chicago
Motor Scooter Division
ip the Following Rangers At Once: Via
State
IMPORTANT

MODEL NO.		PRICE EACH	TOTAL
	RANGER SCOOTER		

▲ **1939 Mead Ranger Order Blank**
The Mead Company of Chicago was so taken with Siegal's creation that it hired the motorscooter mogul to craft badge-engineered versions of the Moto-Scoot as Mead Rangers. Mead had established outlets throughout the country for its long-running line of bicycles, and so became the first large-scale seller of scooters. In the 1930s, nationwide advertising for small companies such as Salsbury was almost impossible, relegating the Motor Glide to being merely a West Coast phenomenon. Larger firms such as Cushman and Mead took out ads in national magazines such as *Popular Mechanics* where the techno-minded could write away for the latest info and an easy order blank for these gee-whiz motorscooters.

▲ **1939 Moto-Scoot**
By 1939, Moto-Scoot—with the aid of its Mead Ranger sales—was indeed "America's Most Popular Motor Scooter," as this brochure boasted. *Time* magazine profiled Norman Siegal as the king of scooters in 1939. Siegal had built 186 Moto-Scoots in his firm's first year, 1936, and was churning out a whopping 4,500 in 1938 with dreams of building—and hopefully selling—more than 10,000 of the midget putt-putts in 1939. Siegal's company had blossomed in a mere three years from a team of three workers handcrafting scooters in the corner of a West Side Chicago warehouse to having its own factory with 75 employees. It was not for naught that *Time* crowned Siegal "the Henry Ford of the scooter business."

Puddlejumpers and More

Following the Leaders All the Way to the Bank

"Dash hither and yon in gay abandon..."
—1936 Salsbury Motor Glide Ad

The design of Salsbury and Elmore's 1936 Motor Glide spread like wildfire. By the end of 1936, every other back-alley garage in America seemed to be a scooter factory, and the back pages of mass-circulation magazines like *Popular Mechanics* were chock full of ads for potboiler scooters. Jukebox maker Rock-Ola had its motorscooter offering that battled for buyers alongside the Powell, Minneapolis, aptly named Puddlejumper, and many more. There was a motorscooter craze sweeping the nation.

The Salsbury Motor Glide had defined the Five Commandments of a motorscooter that set the style for all scooters that were to follow: A small motor placed next to or just in front of the rear wheel; a step-through chassis; bodywork to protect the rider from roadspray and engine grime; small wheels; and an automatic transmission/clutch package (which Salsbury would introduce in 1937). All other scooters that came after the Salsbury had at least three of the five tenets.

▲ **1941 Custer**
Why do Americans love to name motorized vehicles in honor not of fallen heroes but of fallen despots? Consider the DeSoto automobile brand named for Spanish conquistador/Indian slaughterer Hernando de Soto. The fabulous Dodge Coronado named for another conquistador. Or the 2.5hp Custer scooter from Dayton, Ohio, that brings to mind General George Armstrong Custer, the war against the Plains Indians, and Little Bighorn. Perhaps these names should have been indicative of the vehicles' integrity as well. It's one of those academic questions that scholars love to endlessly debate, and we'll leave it in their hands.

▲ 1938 Rock-Ola
Canadian David C. Rockola made his name famous by emblazoning it across the belighted bellies of the jukeboxes his Rock-Ola firm built in Chicago starting after Prohibition's repeal. Rockola was one to go for the gold when the timing was right, and so he next entered the scooter market. Just as jukeboxes were bedazzled with chrome frou-frou and Buck Rogers class, Rock-Ola scooters were stylish little two-wheelers—at least they had compound curves in their rear bodywork, more than you could say for the Cushmans of the day. In 1938, you could choose between the Rock-Ola Deluxe model with the 1hp Johnson "Iron Horse" single-cylinder engine and the Tourist Model RMS-45-38 with 3/4hp. To go, you simply pushed a foot lever to tighten a rubber belt that transmitted power to the rear wheel, saving all of the mess and bother of a clutch. To stop, you pulled the handgrip brake, just like riding a bicycle. The most exciting feature on the Deluxe, however, was the "revolutionary new 'Floating Ride'" suspension on the "rocker spring front frame"—yup, those are screen-door springs on the front forks. Owner: Herb Singe.

▲ 1938 Puddlejumper
When Midget Motors of Kearny, Nebraska, announced its Puddlejumper scooter in 1938, boyish hearts must have done somersaults. Here was Buck Rogers' spaceship come to earth, or at least a motorscooter that could blast off. The styling was radical, far beyond what the neighboring Cushman company was creating, and must have inspired many a scooter stylist (if there was such a thing) to rush back to their drawing boards. Here was a scooter that was actually streamlined—leaving other scooters, cars, coffeemakers, and toasters of the time to cry tears in their beers—and a company that promoted its scooters' "snappy fast performance." Three Puddlejumpers were offered—the Standard, De Luxe, and three-wheeler ("self-balancing"!) Powertrike—with either gas or electric power. But never mind all that. The Puddlejumper even had something more, something no other scooter maker promoted: The Puddlejumper even boasted "Plenty of Leg room."

▲ 1939 Powell Streamliner 40
Channing and Hayward Powell built vehicles to the tune of a different plumber. Based in Los Angeles, the dynamic duo created a long line of scooters from their first 1939 Streamliner through a gaggle of miniature motorcycles in the 1950s, when they switched over to build plywood-bodied pickup trucks that were half a hunter's dream, half James Bond. This first-year Powell Streamliner was powered by a 2.3hp Lauson engine with "forced blast cooling" (a glorified flywheel fan) and centrifugal clutch. The Powell scooter's streamlining was a key to decreased aerodynamic coefficient of drag and optimum handling prowess as it huffed and puffed to reach its top speed of 30mph. Owner: Herb Singe.

"Cheaper Than Shoe Leather"

Scooters Save Soles in the Great Depression

All scooter ads of the late 1930s hailed the go-forever-on-a-teaspoon-of-gas fuel consumption and inexpensive operating costs. This second coming of the motorscooter arrived at the right place at the right time: America during the Great Depression. E. Foster Salsbury had envisioned his scooter being an ideal substitute for an automobile for people who didn't have the money in those bad times. And a Motor Glide was a perfect second car for people who were a little better off. Cushman and others followed this thinking—to the point where Cushman even boasted that operation of its scooters was "Cheaper than shoe leather."

Bad times was the mother of invention for motorscooters. The first scooters of 1915–1920 failed because they were frivolous toys; the Salsbury, Cushman, and Mead–Moto-Scoot scooters were born from economic necessity and succeeded. The mass mobilization to motorscooters in Europe and Japan after World War II would follow this same pattern. And still today, the scooter makers that survive and thrive sell primarily to a market that needs economic transportation.

▲ 1937 Moto-Scoot Challenger
Moto-Scoot quickly created an economy line of scooters—which may have been an oxymoron in the scooter world of 1937—with its new Challenger. Superlatives couldn't begin to describe the Challenger, so Moto-Scoot compared it to the only thing that *could* compare, the Magic Carpet of Bagdad: "Just imagine owning this remarkable little vehicle that gives you a new motion, rivaled only by the mythical 'Magic Carpet of Bagdad.'" And like the magic carpet, the Challenger got great gas mileage—120mpg, which is actually surprisingly low for a 1hp engine that only propelled the scooter to a 30mph top speed. But at a delivered price of $69.50, the Challenger was definitely a deal on wheels if you couldn't own its magic-carpet soulmate. The copywriting was almost intoxicating: "Thrill to the instantaneous response to your throttle...the steady purr of the engine...as it sweeps you to your destination." *Herb Singe Archives*

▲ 1937 Cushman Auto-Glide
Cushman copywriters must have lived simply to extol the economical virtues of the company's great Auto-Glide, but this may have been taking it too far. "Actually 1-10th the operating cost of an auto—yet practically as comfortable and convenient," crows this 1937 flyer. The key word here was "practically," but for Depression-weary Americans, the Auto-Glide was an attractive—if slightly eccentric—alternative. Cushman also liked to plug its gas mileage: "100 Miles on a Gallon of Gas, 500 Miles on a Quart of Oil!" Although with the Auto-Glide's miniaturized gas tank, no mention was made of the number of gas station stops that would have to be made over a 100-mile jaunt. Still, as another Cushman flyer from 1937 stated, "Fast Enough [sic] to get there Quick, Cheap Enough to Afford to Go!" Oh, the joy those Cushman copywriters must have gotten each time they punched out another exclamation point on their typewriters!

► 1938 Rock-Ola

The riches awaiting the daring entrepreneur who signed on as a dealer for a motorscooter line must have been fabulous given the enthusiasm with which scoot builders, such as Rock-Ola in this 1938 flyer, promoted the possibilities. "BE FIRST in this new dealership opportunity" called the ad. The riches may have been merely a glimmer in an entrepreneur's eye, but scooter builders worked overtime at getting a dealership network set up to increase their sales exposure as most people only saw scooters in the magazine ads. Scoot firms advertised the possibilities of creating scooter delivery services, scoot rental agencies, and more. The farsighted investor could jump on the bandwagon to sell "America's Newest Mode of Transportation"; the motorscooter was a miniaturized smoke-spewing gold mine on two wheels—and if you're reading this now you're probably too late.

▲ 1939 Cushman Auto-Glide and Glide-Kar

"Cheaper than shoe leather" was the promise from on high for the economy the Auto-Glide provided, and it was just the thing for those trying to save money to get ahead or merely survive in the decade following the Great Depression. It was also the *raison d'être* of the motorscooter in the minds of visionaries like E. Foster Salsbury and Norman Siegal, so why not tout the scoot's economy—especially in 1939 when no car ad ever mentioned such a foreign concept as gas mileage. Cushman sure did: "Why walk or use an auto when you can Glide for 1/3¢ per mile!" Why indeed! The Auto-Glide's 1.5hp or optional new 2hp engines returned 125 miles per gallon at a top speed of 30mph. But the true question for academics persists: Was it truly cheaper than shoe leather?

"If you want to conserve Gasoline and Tires . . . and save money to invest in War Bonds, you should consider our Motor Scooters. . . . Remember a 'D' Ration book is good for 1 1/2 Gallons a week . . . at 100 miles per gallon that's better than a 'B' Book."
—Floyd Clymer's sales pitch for the Victory Clipper scooter

Build-it-Yourself Scooters

"First Find an Old Washing Machine Motor. . . ."

Too much free time in the garages of suburbia has led to the invention of everything from early scooters to Apple computers. All you needed was a jig saw, some plywood, a couple of baby carriage wheels, an old washing machine motor, and you too could be the neighborhood Wild One.

In the late 1930s, handymen across the country mailed their twenty-five cents in for dubious kits or putt-putt plans advertised in *Popular Mechanics* to be the first on the block to troll down the street on what to the untrained eye appeared to be nothing more than junk on wheels. One person's junk is always another's treasure.

▲ 1949 Zipscoot
Curiously reminiscent of the Cushman Auto-Glide, America's Outstanding Scooter Value could be had for a fraction of the amount of the real thing, assuming that your mechanical skills were up to par. The boasted top speed of 30mph is unlikely, unless on a treacherous downward slope.

Brand new design. All-Steel construction. Front wheel drive, 2-wheel brakes. Balloon Tires, Powerful motor. Two speeds. Big compartment for pkgs. Needed by merchants everywhere for quick, cheap delivery. Write for special introductory offer—Dealers Wanted.
TROTWOOD TRAILERS, INC.
735 Main Street TROTWOOD, OHIO

◄ 1939 Trotwood Motor-Cycle
The attractive, aerodynamic design was enough to awe any customer. The reasoning behind the unusual front-wheel drive is ample storage space in the rear with the added advantage of sure footing on those cold and icy days.

◄ **"This Speedy Scooter Will Assure You Plenty of Fun"**
Back when anything could be built with a little bit of old-fashioned American ingenuity, build-your-own motorscooter blueprints were a staple of *Popular Mechanics* and *Popular Science* magazines. Although they were a poor cousin of factory-made scooters, most people viewed the scooters as the same sort of recycled-washing-machine-motor death traps. *Herb Singe Archives*

CONSTRUCTA-SCOOT

The best motor scooter value in America that easily and quickly assembles from fully finished factory kits.

AT ONLY BICYCLE COST!

$29.50 Less Motor Complete — **$59.50** With Motor Complete

F.O.B. Chicago, Ill. Write for Literature!

RENMOR MFG. CO. Not Inc. **A-2544 So. Wabash Ave., Chicago, Ill.**

BUILD YOUR OWN
MOTOR-SCOOTER

You can build this beautiful, heavy duty—motorized scooter easily. Kits quickly assembled by anyone. All parts furnished. Costs no more than a good bicycle. $29.50 less motor. $59.50 with motor. Write now for complete facts and literature.

RENMOR MFG. CO. Not Inc. **2540 So. Wabash Ave., Chicago, Ill.**

Always write your name and full address clearly when answering advertisements.

▲ **1939 Renmor Constructa-Scoot**
Proof positive that motorscooters indeed came from children's push scooters! Renmor offered this petite putt-putt with or without the tiny $30 motor. After all, if the customer bought the cheap version, dad's lawn mower engine or mom's sewing machine motor could be bolted on and achieve the same break-neck speeds.

Build your own JET ENGINE!

Order these plans today
1. JET PROPELLED BICYCLE, Assemble your own. Photo and instructions, $1.00.
2. HOW TO MAKE EXPERIMENTAL JET ENGINES. Seven sheets drawings with information and instructions $2.95.
3. BOTH OF ABOVE in one order $3.75.
SEND NO MONEY. Order both at once $3.75 C.O.D. in USA plus c.o.d. postage. Send check or Money Order and we pay postage. Get other information too. Rush Order.

J. HOUSTON MAUPIN, Dept. 55, Tipp City, Ohio

SAY YOU SAW IT IN POPULAR MECHANICS

◄ **1951 Jet-Propelled Bicycle**
Although not quite an F-15 fighter plane, the Jet-Propelled Bike was probably as exciting and definitely as dangerous. For just a pittance, Mr. Maupin would send you a few mimeographs, and look out Boeing. From this *Popular Mechanics* ad drawing, the bicycle looks precariously lopsided, so better take out a bloated life insurance policy at the same time.

The Fully Automatic Scooter

Salsbury's Torque Converter Sets the Pace

The earliest motorscooters were hardly mechanical marvels. The engineering was crude, heavy, and only slightly more reliable than a donkey. All that changed in late 1937 when Salsbury announced its new 1938 Motor Glide Models 50 and 60 with Self-Shifting Transmission. The future was here today.

Suddenly the motorscooter ante had been raised to new, dizzying heights, and Cushman, Moto-Scoot, and others who wanted to stay in business rushed to create their own versions of Salsbury's automatic clutch and torque converter. Even today, the son of Salsbury's self-shifter lives on in scooters from Vespa to Honda, and E. Foster Salsbury still receives his royalty check.

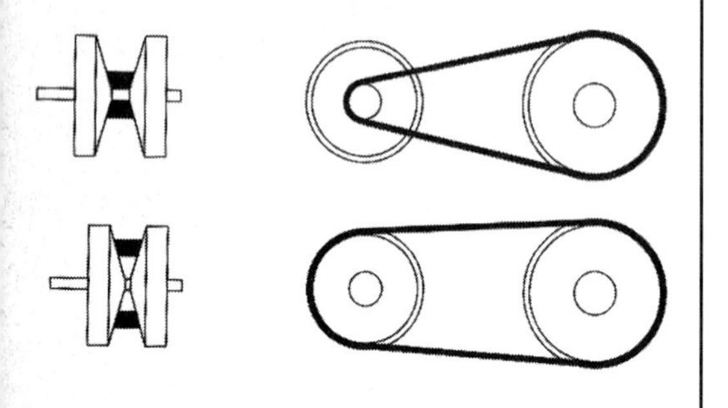

▲ **1937 Salsbury Self-Shifting Transmission**
Buck Rogers technology for the earthbound scooterist: This diagram shows the operation of the Self-Shifting Transmission with the pulley open and contracted for slow and high speed. Secondary drive after the rubber-band primary drive came from a traditional linked chain hitched to a sprocket on the rear axle. This torque converter transmission was far-out in the 1930s, but today seems ho-hum; the reason is due to Salsbury's design, which is still state of the art on industrial engines and vehicles of all sorts, including the latest generation of Honda four-wheelers.

▲ **1939 Salsbury Self-Shifting Transmission**
It was the shot heard 'round the world heralding a new scooter revolution. In late 1937, Salsbury introduced on the new 1938 Motor Glide its most radical feature: the Self-Shifting Transmission, as the firm termed its new automatic clutch and transmission torque converter. Shown here attached to the 8.94ci Lauson side-valve engine, the primary drive was via a V-shaped rubber belt running between pulleys. As speed increased, the spring-loaded driving pulley compressed together creating a larger circumference for the belt to ride: the "gear ratio" could alternate between 14:1 at low speed to 4:1 at top speed—and an "infinite number of ratios" in between, as Salsbury ads like to proclaim. This automatic transmission technology was startlingly innovative in 1937, and every scooter maker that wanted to survive created its own version of the Self-Shifting Transmission. *E. Foster Salsbury Archives*

► **1939 Salsbury Motor Glide Model 72 and Acrobat**
Look, no hands! Riding the new Salsbury scooter with the radical Self-Shifting Transmission was so simple, "one of Hollywood's loveliest ladies," as this 1939 press release and photo lauded actress Mildred Coles, could scoot along with her feet doing the driving. Salsbury ads and promotional stunts showing women on Motor Glides were more than just cheesecake, as the scooters were aimed at a market looking for an economical vehicle that was easy to use. Never one to rest on his laurels, E. Foster Salsbury introduced two all-new Motor Glides in 1939 with the Self-Shifting Transmission—these marked the seventh and eighth models that the firm created since the first Motor Glide all the way back in 1936. The torque converter first appeared in late 1937 on the 1938 Model 50 with 5/8hp and the Model 60 with 1–1.5hp Johnson engines. The revamped Models 70 and 72 both used a new 2.3–2.7hp Lauson single-cylinder four-stroke measuring a square 2 1/4x2 1/4in for 8.94ci—marking the third shift in engine manufacturers since 1936 for the firm. The Self-Shifting Transmission was standard on the Model 72 and optional on the 70. The most dramatic change was the new "aerodynamic aircraft designing for style leadership," as Salsbury brochures waxed poetically. These radical styling advances were most pronounced in the new Easy-Lift Streamlined Engine Hood, which truly looked as though it was "styled by the wind" with curves replacing the creases and folds of the older Motor Glides. The sides of the hood were now made of mesh, providing a tantalizing and seductive see-through view of the powerplant and radical transmission in action. This was the future in scooters, and it was definitely as new as tomorrow.

"The 1938 Motor Glide is the last word in personal transportation and far exceeds my fondest expectation for performance."
—Air racer and barnstormer Colonel Roscoe Turner

The Chic Motorscooter

Styling Comes To The Putt-Putt

And then in the late 1930s, motorcycle tuner, racer, and builder Albert G. Crocker created his Scootabout scooter, also in California. The Scootabout's engine and solid front suspension were typical of the time, but not so its flowing bodywork. Crocker followed the tenets of Salsbury's mechanical design but added the Sixth Commandment of a scooter: styling.

▲ Bathing Beauty and 1939 Crocker Scootabout
Beauty came to the scooter beast in the late 1930s when Albert G. Crocker created his Scootabout. The mechanicals of the Crocker were typical of the time, but not so its flowing bodywork. The Motor Glides and Auto-Glides initially did not venture beyond single-plane folds in their bodywork—after all, scooters were built on a budget and sold as a utilitarian device.

◄ Happy Scooterist and 1939 Crocker Scootabout
The Crocker was formed of sinuous bodywork with two-tone paint that added a further dimension to the curvaceous lines. The tail section covered the wheels with fender skirts, the design coming together at the rear in a teardrop shape. With its Streamline Moderne teardrop shape and flashy Art Deco paint scheme, the Scootabout foreshadowed the spirit of the Jet-Age scooter styling that was to come.

▲ 1939 Crocker Scootabout
A study in styling contrasts. Underneath the flowing bodywork, the Crocker had advanced little from the Motor Glide: It featured a rigid chassis without suspension (springs were added in 1941), an automatic clutch, and a brake on the rear wheel. But the streamlined engine cover inspired scooters of the future by adding complex curves to the metalwork, and suddenly that little scooter was cute—a great sales point! The two-tone paint scheme featured a rich black base with red scallops highlighted by gold pinstriping. As a later Crocker flyer shouted, the Scootabout offered "New Luxury in motor scooters." Owner: Herb Singe.

► 1941 Crocker Scootabout
This 1941 flyer signified a meeting of two great minds: Albert G. Crocker and motorcycle salesman, promoter, and all-around kingpin Floyd Clymer. Crocker was a former colleague of Harley–Davidson's famed engineer William Ottaway when the duo worked for Thor motorcycles; Crocker was later an Indian dealer and racer, so he was well placed to be a thorn in everyone's side. Crocker also built the great Crocker V-twin motorcycles of the 1930s that single-handedly caused Harley–Davidson and Indian more grief than any event up to the British motorcycle invasion of the 1950s. Clymer, meanwhile, was an ambassador for motorcycling throughout the twentieth century, with his hands in more aspects of cycling than anyone could count, including Clymer Motors of Los Angeles and his later cycle magazines. When Clymer decided he needed a scooter to sell during the craze, he called up Crocker, who created this machine. But the Scootabout never scooted about much: production probably numbered less than 100—as did production of Crocker's infamous motorcycle.

The Quaint Sport of Scootering

Over Hill, Over Dale, the Scooter Goes Everywhere

Like wildfire, the scooter sparked a craze, and people took to doing daffy things aboard their newfound friend. Those Hollywood types took to tearing up the town on their Salsbury scoots, while across the country scooter rental shops allowed anyone with a couple pennies to terrorize traffic on a putt-putt.

It was the birth of the quaint sport of scootering.

▸Moto-Scoot Goes Uptown
In the glorious days of Pierce Arrows and Duesenbergs, what could have induced more side-splitting hoots of laughter than even the notion of someone dashing to the opera on a Moto-Scoot? (They must be nouveau riche!) This 1939 *Collier's* magazine cover was probably created half in jest at the blossoming scooter craze and half in fear that such a sighting could actually occur. But what a way to go: Leave the chauffeur at home to raid the fridge and save pennies on gas all at the same time. Note that Milady is not sitting side-saddle but instead showing a bit of racy leg as the dynamic duo arrive at the theater, wave off the valet, and find the best parking spot nestled betwixt two Packards. Definitely nouveau riche. *Herb Singe Collection*

◂The Modern Woman and 1936 Moto-Scoot
By the late 1930s, the scootering craze was all the rage, and Chesterfield cigarettes promoted an image of the modern woman dashing about on her own two-wheeler and daring to openly smoke a cigarette. Think of this 1937 magazine ad as a precursor of the 1970s Virginia Slims cigarette ads with their slogan, "You've come a long way, baby!" The Moto-Scoot also promoted the image of the free-wheeling woman freed from earthly chains and duties, which had indeed been one of the first sales pitches for scooters—not only were they easy to operate, but you didn't have to sit side-saddle anymore. The look of astonishment on the male hitchhiker's face undoubtedly turned to disappointment as the woman, her two-wheeler, and her pack of cigs disappeared over the next hill, leaving him with lipstick traces, dust, and cigarette smoke. *Herb Singe Collection*

Collier's
October 7, 1939
5¢ A COPY
WEEKLY
CORONADO FEVER
A Short Story by Frederick Hazlitt B

High Tech Scooters, Circa 1930s

The Sound and the Fury

Tomorrow and tomorrow did not always arrive when you rode a motorscooter in the 1930s. Marvel at their motors, but mechanical marvels they weren't.

Still, scooters were a fun fad, a quaint little vision of the future, and all around, their mechanical features were probably on par with the cars of the time.

▲ Streamlining and the 1938 Moto-Scoot
Streamlining was all the rage in the 1930s. The concept of speed had grasped the world through the heroic feats of pioneering aviators, and streamlining was a result of a new hybrid of science and engineering: aerodynamics. The Streamline Moderne school of architecture and industrial design was quickly adapted to everything from toasters to locomotives, diners to motorscooters. Needless to say, aerodynamics were of the utmost importance at scooter top speeds of a breathtaking 30mph, as with this 1938 Moto-Scoot. According to this ad (see the top left mini-photo), "streamlining" seemed to mean that the scooter was narrow in width. But the new metal wicker-pattern engine cowling must have played a part as well, with its styling that was half Streamline Moderne, half wastebasket on wheels. Nevertheless, the aerodynamic efficiency of the new cover must have been fabulous—witness the speed streaks captured on film on the cover of this rare brochure. But look closely, and you may agree that this woman's cape was being held aloft by a helping hand that was later airbrushed out. *Herb Singe Archives*

▲ 1939 LeJay Electric Rocket
The gasoline-fueled internal-combustion engine was not the only motive power to propel motorscooters. By mailing in your hard-earned 25¢ to the LeJay Manufacturing Company in beautiful Minneapolis, you could get "complete simplified plans" to build your own LeJay Electric Rocket. As this 1939 mail-order ad from the back pages of *Popular Mechanics* boasted, the Rocket had no transmission and no clutch, and was suitable for sidewalk use.

◄ "Suspension" and the 1937 Auto-Glide Model 1
Suspension—or the lack thereof—was one of the major arenas of battle on the scooter technology front. Protecting the rider's gluteus maximus paid dividends in scooter sales, so manufacturers rushed to add "suspension" to their hitherto rigid-framed wares. This 1937 Auto-Glide Model 1 was typical of the earliest scooters and proves that riding a pioneer scooter was not for the faint of heart: front suspension was provided sheerly by the hard-rubber tire and the inherent flex and compression of the solid metal forks. Before makers such as Moto-Scoot bolted on sprung bicycle seats to save wear and tear on a rider's behind, that behind provided most of the suspending. By about 1939, most makers had added some sort of crude lever-action rig to the front forks sprung by a hardware store-variety screen-door spring; later "advances" used barrel springs, which provided a "progressive-rate" spring, of sorts. Rear suspension came into vogue about the same time, usually by means of two barrel springs. Still, despite all the advances, American motorscooters—even into the 1960s—were never renowned for the quality of ride.

▲ 1937 Salsbury Aero Model Motor Glide with Cycletow
The nifty Cycletow attachment was like a set of training wheels for your Salsbury, turning it into a four-wheeler. Designed for car dealerships such as Felix Super Chevrolet—"Where Friend Meets Friend"—the idea was that your teenage go-fer could jockey the Motor Glide across town to pick up an ailing auto, then pilot the car back towing the cycle. The concept received widespread use from the 1930s through the 1960s as many scooter and motorcycle makers hawked similar options: Cushman had its Trail-It (which reversed the scoot's direction for towing), and Harley-Davidson and Indian had the three-wheeled Servi-Car and DispatchTow, respectively.

▲ 1939 Cushman Auto-Glide Model 2 with Side-Kar
With the addition of what looked like a miniature metal coffin on wheels, the Auto-Glide could be turned into a family troop carrier capable of hauling pop, mom, and little Suzy. Salsbury, Moto-Scoot, and Mead also offered a sidecar to transform their scooters into precursors of the family station wagon. By 1939, Cushman had added its Model 2 to the line-up complete with a new 2hp engine, ideal for the rigors of Side-Kar toting. This early Cushman flyer noted that the "extra wheel of the Side-Kar pleases people who do not care for the 'balance' of two wheels."

Waxing Poetic on Scooter Woes

The true unsung heroes in the scooter engineering epic were the ad writers. Creativity, exaggeration, and exclamation points coursed through their veins like gasoline through their scooters' engines. They waxed poetic to accentuate the positive and hide the many flaws, creating putt-putt poems in what to the uninitiated was mere ad copy. It was a Herculean task if ever there was one. Herewith, consider some of their finest phrases:

- "Streamlining" was the perfect description for refrigerator-like styling from Salsbury to Crocker to today.
- "Floating Ride" was Rock-Ola's term for suspension made up of two screen-door hardware-store-variety springs.
- "Sane Speed" was Cushman's way of promoting lack of speed: "FAST enough to get there quickly and SLOW enough to be absolutely safe."
- "Forced Blast Cooling" was the Powell Streamliner's inflated promotion for its flywheel fan, which appeared on most every scooter on the market.
- "Self-Balancing" ballyhooed anything with more than two wheels, as with the three-wheeled Puddlejumper Powertrike or Cushman's Side Kar.
- "Iron Horse" was Rock-Ola's name for its Johnson 1hp engine, likening it to a steam locomotive.
- "Plenty of Leg Room"—only the Puddlejumper scooter had it!

Chapter 3

MILITARY MOTORSCOOTERS

1939-1945

Secret weapons win wars. Vikings had their long ships. Trojans had their horse. Hannibal had his elephants. Conquistadors had their galleons. And in World War II, the Allies had paratrooper scooters.

Dropping scooters out of the sky began as a European pastime. It's rumored that the Nazis created a miniature motorscooter to accompany the *fallschirmjäger* on invasions, but no photos or descriptions of such a scooter have surfaced. Near the start of the war, meanwhile, the Italians crafted the Aeromoto, a micro-sized four-wheeled paratrooper putt-putt built by Società Volugrafo of Turin.

It was the British and Americans, however, that perfected the paratrooper scooter: the British built their Welbike while the Americans had the Cushman and Cooper. Development of the paratrooper scoots was top secret during the war, but spies were actually more interested in the big bang of the Manhattan Project.

During World War II, scooters were shanghaied into service on other fronts as well, fighting for peace, dignity, and the American way with all the nationalistic fervor the elfin engines could muster. Scoots were drafted into the military to serve as base and airfield go-fers.

On the home front, scooters fought in civvies; most U.S. car makers were converted solely to military contracting, so the government granted special dispensations to some US scooter builders to build scooters as civilian wartime transportation. Riding a scooter became a wartime duty, as patriotic as flying the flag or planting a victory garden.

The motorscooter may not have won the war, but it was certainly the mouse that roared.

▸ **Scooters at War**
Motorscooters served their country at war and on the homefront, where gas rationing made prewar scooters the ideal transportation. At war, they made decent shields to hide behind from bullets. The ability to have instantly mobile troops behind enemy lines, however, was no laughing matter. Concealing their location, on the other hand, could be a little difficult with a puttering two-stroke and a trail of blue smoke.

Meco WELDING EQUIPMENT
FULLY EQUIPPED
UNBLENDED
PRIME MALT
A.P.A. TUNING
A FREQUENCY M.C.
VOLTS
200 CARTRIDGE
7.62 MM
CARTONS,
LOT LC L-126

Paratrooper Putt-Putts

Scooters From Heaven

The strategy was simple. Paratroopers would drop in the dead of night deep behind enemy lines, unpack their elfin scooters, and lead a true charge of the motorized light brigade on putt-putts. Nothing would stand in their way.

The logic was sound, but the means left something to be desired. These scooters never went into wide-scale use and ended their service to be sold off as war surplus, finding a welcome home with today's scooter collectors.

World War II could have ended differently if scooters had been allowed to fight for their country—the Allies could have lost.

▸1944 Cushman Model 53 Airborne Scooter
Mussolini had his Aeromoto, Churchill had his Welbike, and Roosevelt had his Cushman. As part of the plan for the 82nd and 101st Airborne to land in Normandy and zip across Europe to Berlin, the Model 53 was tested behind the Cushman factory by tossing a rope over a tree branch, lugging the lean green machine into the air, and letting it drop; when it no longer broke, it was ready for action. Extras such as suspension and headlights were deemed luxury items, but the chassis was reinforced and bracing was added around the engine awaiting the big fall from the heavens.

◂1944
Cooper War
Combat Motor Scooter
While Hitler was shooting V1 rockets at Britain, American mad scientists were testing a new secret weapon, the Cooper paratrooper scooter. In the midst of World War II, the US government scoured the country for the best scooter manufacturer and came across a couple of winners, Cushman and Cooper. The Cooper was a Powell A-V-8 scooter that Frank Cooper decorated with his own decals and reinforcing bars.

"World War II is responsible. Sprawling war plants and the mobility-conscious armed services found a myriad of uses for the powered runabouts, which previously had been notable chiefly as a special headache to traffic safety planners."

▲ Lambretta Fold-Up Paratrooper Scooter
Inspired by the Volugrafo Aeromoto, Innocenti created this prototype paratrooper scooter. The engine was based on the Model C 125cc engine, without all the frills of covering the scooter with fancy bodywork. After all, this was war.

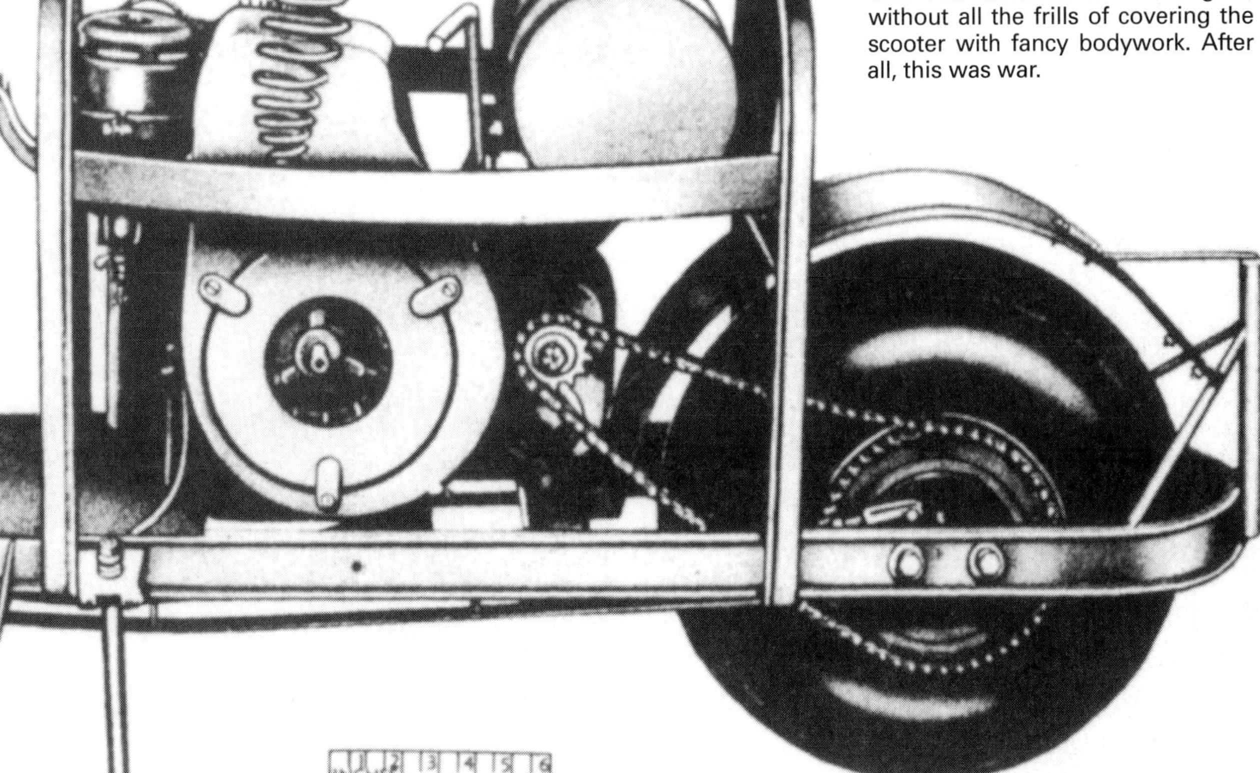

◄ 1944 Cushman Model 53 Airborne Scooter
This 255lb Cushman is ready for the fall. This restored Model 53 is equipped complete with parachute and GI helmet. Never sent into action, many Model 53s were repainted a more appropriate yellow for military duties on the home front. The 6.00x6in tires with the 4hp Husky engine reportedly could carry a passenger at 40mph with a maximum payload of 250lb. Owners: Keith and Kim Weeks.

► Volugrafo Aeromoto Paratrooper Scooter
This scooter could sure do some damage if dropped 5,000 feet onto the enemy. The average soldier could carry the Volugrafo around just like a 50lb suitcase, or drive the 125cc two-stroke over nearly any terrain thanks to the two-speed gearbox. Società Volugrafo of Turin manufactured the Aeromoto, which subsequently inspired an arms race as the Allies rushed to create their own paratrooper scooters.

Putt-Putt Attack!!

Top-Secret Scooters

The military motorscooter was conceived to make Hitler's Blitzkrieg look positively sluggish. Scooters were to give individual mobility for all troops as they revved their miniature motors in bloodthirsty patriotism. Despite numerous attempts to create military motorscooters, however, the patriotic putt-putts remain pacifists.

Perhaps the idea was wrong. Perhaps scooters could have been used to create instant foxholes when dropped from 1,000ft. Or simply just dropped *on* the enemy. Or perhaps their true story has never been told; perhaps they served their country as the ultimate counter-espionage decoy, when plans to use them were allowed to fall into enemy hands to divert attention from the Manhattan Project.

▲ Cushman Auto-Glide 34 with Side Kar
"This was a tremendous thing. We were making scooters when Ford couldn't get tires to make cars," Robert H. Ammon recounted to the *Lincoln Star* about his scooter production. Thanks to a small military contract of 495 Model 34s with side cars, Cushman Motor Works was able to keep pumping out their putt-putts between 1942 and 1945. The War Department decided that scooters were economical civilian transportation so Cushman was still able to produce up to 300 scooters a day during the war.

◄ Administrative Cushman Auto-Glide 34
The Auto-Glide 34 with sidecar was not used in combat but rather to wheel soldiers around the base. Nevertheless, drab military green was a must, as opposed to the earlier bright colors to lure the public to reach for their pocketbooks. This was the last model to carry the name "Auto-Glide." Cushman founder Robert Ammon remembered, "That's when they really took off, right after World War II."

► Welbike Battles for Britain
The British Welbike parascooter could handle any terrain as these adept scooterists demonstrate. Perhaps the noise and exhaust of the two-stroke, 98cc Excelsior engine putsing along with its single-speed gearbox wouldn't allow for a good ambush, but at least they would make an amusing target for bored snipers. Postwar, the Welbike was domesticated and renamed the Corgi, the same breed as the Queen's famous dogs, and manufactured by Brockhouse of Southport.

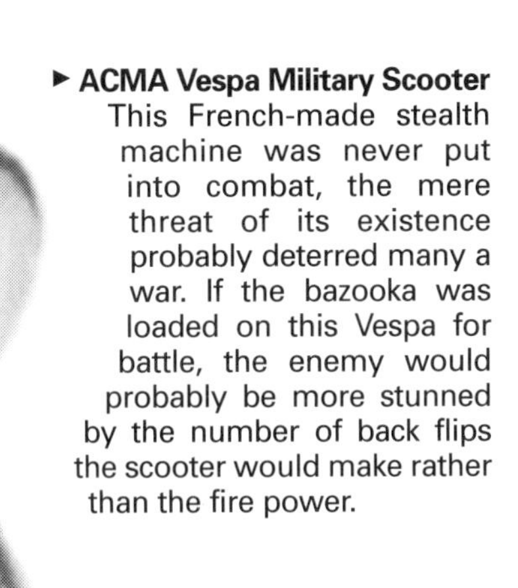

► ACMA Vespa Military Scooter
This French-made stealth machine was never put into combat, the mere threat of its existence probably deterred many a war. If the bazooka was loaded on this Vespa for battle, the enemy would probably be more stunned by the number of back flips the scooter would make rather than the fire power.

"The scooters were drafted into our war plants to deliver small parts between buildings. Sidecars were added to convert them into mobile soda fountains for carrying refreshments to production-line workers. Other sidecar combinations were equipped for ambulance duty and fire-fighting."
—*Popular Mechanics*, 1947

CHAPTER 4

The GOLDEN AGE OF THE MOTORSCOOTER

1946-1954

Bad times begat motorscooters. As the dust cleared away in the days following the end of World War II, scooter makers blossomed throughout Europe and Japan. Just as in the dark years of the Great Depression in the United States, scooters were suddenly seen as the Platonic ideal in cheap transportation to mobilize the masses. If German philosopher Immanuel Kant had written a treatise on motorscooters, and scooter scholars do not believe he did, his categorical imperative would have been the perfect scooting sales pitch: The greatest good for the greatest number.

In the early postwar years, the world rebuilt itself, and "The New" burst on the scene from all fronts. No idea was too outrageous, no invention too far flung, and science would save us all, as many believed it had done in creating the atomic bomb and ending the war.

"The New" was everywhere. The bikini bathing suit, created by French *coutourier* Louis Reard, was first modeled by a stripper at a Paris fashion show in 1946. In California in 1950, Fender introduced its first mass-produced solid-body electric guitar, the Esquire, dubbed by many as "an electric canoe paddle." In Japan in 1952, Sony launched the first pocket-sized transistor radio. In the US of A in 1954, Swanson Frozen Foods debuted the TV Dinner.

In both Italy and Japan, the first scooters hit the road in 1946; France, Germany, England, and others had their own makers within several years, most often licensed versions of the Italian makes, from Lambretta to Parilla.

This third wave of scootermania shaped a further clause in the definition of a scooter: Motorscooters were born from economic necessity. The first scooters of the 1910s and 1920s had been gadabout toys, and failed. The second wave of scooters, sparked in the United States, was a byproduct of the Great Depression.

Postwar, the European and Japanese economic recovery rode atop the midget wheels of motorscooters, as well as other mopeds, commuter motorcycles, and microcars. Cushman even helped establish the Belgian Cushman firm under the auspices of the U.S. Recovery Act.

It was the dawn of the Golden Age of the Motorscooter.

"Just like Henry Ford put the workers on wheels in America, we put automotive transport within the reach of people who never expected to travel that way."
—Enrico Piaggio, *Newsweek*, 1956

▶**Golden Memorabilia from the Golden Age**
Following World War II, motorscooters appeared everywhere like flowers in the spring. Italy and Japan developed and released new scooters almost simultaneously in 1946, and the rest of Europe was quick to follow. The U.S. makers merely pushed prewar designs back into business. The Golden Age of the scooter was at hand.

MOTO
PARILLA
VESPA CLUB d'ITALIA
ASSO
MOTO PARILLA
IMPERIAL SERIES
SALSBURY
Lambretta
THE REVOLUTIONARY
SALSBURY
SUPER-SCOOTER

Birth of Piaggio's Vespa

The Scooter that Saved the World

Società Anonima Piaggio was founded in 1884 by Rinaldo Piaggio in Genoa, Italy, to make woodworking machinery and, later, railroad cars. In 1915, Piaggio delved into aviation, inventing such innovations as cabin pressurization. In the waning years of the war, the Piaggio factory at Pontedera was subjected to urban renewal courtesy of B-17s; the surviving machine tools were confiscated by the Nazis. The factory was rebuilt after the end of hostilities, and Piaggio began looking for a new product to sell.

Development of the Piaggio Vespa began in autumn 1945, just months after the liberation of northern Italy; the first Vespa was for sale in April 1946. From its beginning, the Vespa revolutionized the motorscooter, and from that day forward it became the standard—even the Platonic ideal—by which all other scooters the world over were judged.

▲ 1946 Vespa Sales Flyer
Piaggio named its scooter the Vespa not for the ancient Roman emperor Titus Flavius Sabinis Vespasianus, but rather for the buzzing of its two-stroke engine that sounds like a wasp, or *vespa* in Italian. The styling of the scooter's tail also bore an odd metallic resemblance to a wasp's abdomen. But Piaggio's Vespa was not the only Vespa around: Rival makers MV Agusta and Moto Rumi also offered motorcycles named Vespa, although as soon as Piaggio's scooter caught on they were lost in the shadows of *the* Vespa's limelight. Throughout the world, scooter became synonymous with Vespa and the words are almost interchangeable. In 1946, the first year of production, 2,484 Vespas were built; in 1947, production skyrocketed to 10,535. By 1994, Piaggio had built more than 10 million Vespas. It all began with this advertising brochure, one of the first Vespa sales flyers.

◄ 1945 Paperino Prototype
The Vespa was almost called the "Donald Duck" scooter, although its doubtful Piaggio ever applied for a trademark license from Walt Disney. Fiat called its 500cc midget car the Topolino, or "Mickey Mouse," so it seemed only fitting that Piaggio's first scooter prototype was named Paperino, Italian for the cartoon quacker. Both vehicular monikers traded on the Italian love for Disney characters, which are nearly as popular as images of the Madonna in Italy. But it was the styling that doomed the Paperino. "Admittedly," Enrico Piaggio later said in retrospect, "the first motor scooter was a horrible-looking thing, and people ridiculed us to our faces." After creating the single Paperino prototype, d'Ascanio returned to his drawing board.

◄ 1946 V.98 Prototype
Legends abound concerning the early Vespa prototypes. The wheels and their stub-axle mounts are rumored to come from leftover World War II bomber landing gear. The engine is believed to have been a starter engine for Piaggio wartime airplanes. And Enrico Piaggio was in such a hurry to get the Vespa to market that there was no time for full testing of the prototypes, so they were driven on dirt roads without air filters to facilitate wear and tear on the internal organs. The V.98 prototype was first shown to the public at the 1946 Turin Show, and 100 pre-production prototypes were built before the assembly lines started rolling.

The Salsbury Connection

In 1938, E. Foster Salsbury sent a sales agent to Europe armed with top-secret blueprints and photos to discuss licensing Salsbury scooter production to potential scooter makers. The agent met with many firms throughout England and the continent; E. Foster Salsbury said in a 1992 interview that, while he himself was not on the licensing trip, he believed that Piaggio was one of the firms approached. No records of the trip remain in Salsbury's archives, so the scootering world may never know for sure the provenance of a Salsbury-Vespa lineage. The one thing known for certain is that in those days with a second World War looming on the horizon, no European firms actually signed on the dotted line to build a licensed version of the Salsbury.

▲ 1948 Vespa 125
Your ticket to *la dolce vita*. The original Vespa of 1946 was a bare-bones, no-frills scooter: It had a single saddle seat and was powered by a 98cc engine that created a puny 3.3hp and propelled the scooter to only 35mph. With the introduction of Innocenti's Lambretta in 1947, Piaggio was forced to refine its model. The Lambretta came with two seats, marking it in people's minds as a more "social" scooter in the best Italian sense of the word, and was powered by a 125cc engine from its birth. By 1948, Piaggio matched the Lambretta by adding a second seat for side-saddle riders and creating a 125cc engine by increasing the bore from 50x50mm to 56.5x49.8mm, which pumped up power to 5.5hp and top speed to 44mph. In addition, the front and rear suspension was redesigned to give a smoother ride and combat the old problem of braking squat. And riding on its wave of public acclaim, the 100,000th Vespa was built in 1950. *Collezione Vittorio Tessera*

Early Development of the Vespa

The Best Scooter Money Could Buy

◄ **Papal Blessing for the Millionth Vespa** In April 1956, the 1 millionth Vespa rolled off the assembly line, and the occasion was marked by (surely spontaneous) celebrations throughout Europe. The million-mark was tallied from combining worldwide Vespa production; the Piaggio factory in Italy now manufactured 500 scooters a day, and factories in France, England, Germany, and elsewhere were building Vespas under license. A celebration was held in Pontedera, and the Pope was there to give his beatific blessing on the scooter that had provided the wheels for Italy's reconstruction. Vespa Day was declared throughout Italy with festivities held in fifteen Italian cities, including a convoy of 2,000 Vespas traveling en masse through Rome and halting all traffic. It was all a public relations coup for Piaggio, seconded certainly only by the upcoming millionth Lambretta, which would roll off of the Lambrate line in the late 1950s.

◄ **1954 Vespa 125 and *La Bella Donna*** In 1954, Piaggio began a new advertising campaign with annual calendars picturing Vespas straddled by *la bella donna.* It was sex appeal in two senses of the word: The drawings of women attracted male buyers, but they also promoted the idea of owning a Vespa to women as the scooter was so easy to operate *and* was a beauty accessory along with lipstick and high heels. This image came from a 1954 Piaggio calendar featuring Vespas around the world; here in Paris, your typical Vespa rider stops for a makeover before going to church at Notre Dame.

Vespa quickly became synonymous with scooter, and to many people the words are still interchangeable. Since its design was a watershed in scooter technology, many people believe it to be the first scooter ever built. Not so. In fact, the Vespa was not even the first Italian postwar scooter; the Gianca Nibbio wears that crown, and there were also early scooter prototypes built by Fiat. Instead, the Vespa was more important than being simply the first; the Vespa was the best.

Piaggio's Vespa was a scooter a grown person could ride with dignity to work or on vacation, and so the Vespa spread the good word about scooters throughout the world. The Vespa was reliable. It was economical. It protected its rider from roadspray and motor oil. Its superior engineering demanded the respect of even motorcyclists. As a Calabrian scooterist told *American Mercury* magazine in 1957, "Wherever donkeys go the Vespa goes too."

In the end, the Vespa went miles farther than the donkey.

▸ 1951 "Vespone"
By 1951, the Vespa was the pre-eminent motorscooter the world over, already a legend beyond its miniscule size. This *Vespone*, or giant Vespa, model was a masterpiece of overstatement for the 1951 Fiera di Milano. It must have towered some 15ft in the air, complete with monster kickstarter and massive Pirelli tires with whitewalls. It was only the beginning for the motorscooter as icon.

Vespa Creator Corradino d'Ascanio Speaks

D'Ascanio's Vespa was truly revolutionary, incorporating features of motorcycles (two wheels, easy-to-use handlebar-mounted controls, and saddle seats), airplanes (monocoque unit design and single-sided stub axles), and automobiles (protective bodywork, covered motor, and floorboards). In an interview with Italian RAI television, d'Ascanio told of his inspiration in creating the Vespa:

"Naturally at the beginning of the project I was terribly enthusiastic about the idea of designing something new. I did everything I could to make a critical assessment of all existing vehicles and spent sleepless nights trying to understand the project in the clearest, simplest terms.

"Then one day some ideas came to me and I sat down at the drawing board and tried to define the problem like this: I began by drawing a person sitting comfortably on a chair. I drew two wheels symmetrically in front of and behind the person, a mudguard running down from the seat and covering the wheels, and handlebars above the front wheels. Automatically, the Vespa took shape."

D'Ascanio also outlined his parameters in creating the scooter to an Italian magazine: "Having seen motorcyclists stuck at the side of the road many times with a punctured tire, I decided that one of the most important things to solve was that a flat should no longer be a large problem just like it wasn't for automobiles.

"Another problem to resolve was that of simplifying the steering, especially in city driving. To help this, the control of the gearshifting was placed on the handlebars for easy shifting without abandoning maneuverability, making its use intuitive for the novice.

"Another large inconvenience with traditional motorcycles was oil spraying on clothes, so I thought of moving the engine far from the 'pilot,' covering it with a fairing, and abolishing the open chain with a cover placing the wheel right next to the gearchange.

"Some solutions came from aeronautical technology, with which Piaggio was obviously familiar, such as the rear tubular wheel holder borrowed directly from the undercarriage of airplanes. The single shell frame surpassed even the most modern automobile design since the stamped bodywork of strengthened steel was a rarity."

▴ 1958 Vespa 150 and Bathing Beauty
The Vespa was refined annually, both mechanically and physically. The early Vespas boasted a rounded voluptuousness in keeping with the robust Marilyn Monroe image of the early 1950s. By 1958, the styling was facelifted, resulting in a sleeker, more bare-bones scooter that matched the Twiggy lines that were the coming rage. In October 1958, the 150 also received an improved engine with 5.5hp at 5000rpm. The square 57x57mm engine wasn't as fast as the GS—it topped out at 51mph—but got better gas mileage and cost considerably less. By 1961, the standard Vespa 150 had four speeds and was tuned to 6.9hp at 5000rpm for a 56mph top speed and 100mpg.

Innocenti's Lambretta is Born

To the Tune of a Different Plumber

Just months on the treads of the Vespa came Innocenti's Lambretta. The first Lambretta was as different from the Vespa as you would expect of a scooter created by a plumber versus an aviation engineer. Whereas Piaggio's d'Ascanio used a sheet-metal monocoque frame-body unit, Innocenti went with what its designers knew best—and had factories filled with—pipes.

Ferdinando Innocenti was a plumber before starting his company in 1931 to produce steel plumbing pipes and tubing. During World War II, Innocenti produced artillery shells and pontoon bridges, but following the Fascists' defeat, Innocenti and his general director, Giuseppe Lauro, looked for a way to turn swords into plowshares. They came upon the idea of building a scooter that all of Italy could afford to own and operate.

"Let the wife use the family car. Go to work on the amazing Lambretta."
—1950s Innocenti ad

In 1945–1946, the duo assigned engineer Pierluigi Torre to create a scooter, which they named the Lambretta after the factory's site on the Lambro river in the Lambrate quarter of Milan.

The New Lambretta Poster
A "new" Lambretta arrived almost every other year—as did a "new" Vespa. It was tit for tat in an ongoing war between Innocenti and Piaggio. When the Lambretta had a 125cc engine, the Vespa's 98cc powerplant was junked to be replaced by a 125cc. A similar scenario took place with larger wheels, twist-grip shifters, dual seats, suspension upgrades, and more. It was a war in the time-honored tradition of the Romans and Visigoths, Montagues and Capulets, and everyone within the Italian government.

▶ 1949 Lambretta Model B and Happy Lass
This lass was happy to have an updated Model B Lambretta. Innocenti's Model A was not as grand a hit as the firm had hoped for, partly because buyers lacked confidence in the elfin 3.50x7in tires. Thus, in December 1948, Innocenti created a revised version of the A, called, naturally enough, the B. This new model rode on larger, 3.50x8in wheels, featured a left-hand twistgrip gearchange using a push-pull cable system rather than a foot lever. With the B, Innocenti had a hit. *Collezione Vittorio Tessera*

▶ 1947 Lambretta Model A and *Lo Cowboy*
The launch of the Lambretta was preceded by months of advertising to ready the market for Innocenti's radical new vehicle. Day and night throughout 1947, Italy's RAI radio played a Lambretta commercial with a ditty that became the Shave-and-a-haircut-two-bits of Italy—one of those darn tunes that you couldn't get out of your head and is still subconsciously whistled by Italians *in giro*. One of the first sales flyers for the Lambretta featured this young *vaccaro* in full Wild West regalia with his trusty terrier, illustrating one of the Lambretta's most important features: a second seat for carrying a passenger, marking it in buyers' minds as a "social" scooter rather than the purely functional, one-seated 1946 Vespa. The front saddle seat sat above the gas tank with the rear pillion perched above the toolbox. As part of the never-ending Cowboy-Indian wars between the Vespa and Lambretta, Piaggio would soon counter by adding

1948 Lambretta Model A
Innocenti's first Lambretta Model A was as revolutionary as Piaggio's Vespa, but in a different way. The Vespa was technologically advanced, whereas the Lambretta was spot-on in its utilitarian design, the perfect machine to provide cheap transportation for the masses in war-torn Italy. The Lambretta was more traditional with motorcycle-like features: The chassis was a traditional tube frame, not surprising coming from Ferdinando Innocenti, a former plumber-turned-pipe manufacturer. The Model A lacked protective bodywork and had only scanty legshields compared to its chief rival, but what it fell short of in covering, it made up for with a larger 125cc engine compared to its 98cc Vespa counterpart. The two-stroke, single-cylinder measured 52x58mm with its upright cylinder inclined slightly forward. Power was 4.3hp with a top speed of 40–44mph. Drive to the rear wheel was via an enclosed shaft with bevel gears turning the axle. A foot-operated rocker pedal controlled the three-speed gearbox with an indicator on the inside right footboard. The first Vespa went on sale in April 1946; by October 1947, the first Model A Lambretta rolled off the assembly lines and the race was on. Within a year, Innocenti would manufacture 9,669 of its Model A. Owner: Vittorio Tessera.

Early Development of the Lambretta

Win Friends and Influence People

The Lambretta was ultimately no less influential or important than the Vespa, yet it remains even today in the shadow of its rival—except among most Mod cognoscenti and the masses of India, the latter being no small crowd.

Consider its (ongoing) history: The Lambretta spawned more imitators than the Vespa, as its tube frame with sheet-metal bodywork was easier to copycat. These imitators included several long-lived license-built versions, such as the German *wünderkind* NSU, the Spanish Serveta, and the Indian-built Lambretta, which is still mass-produced, despite being decades out of date.

◄ 1950 Lambretta Model LC Poster
The new C was based on a completely redesigned chassis fitted with the tried-and-true 4.3hp shaft-drive engine. The A/B chassis with its twin-tube construction was replaced by a single large-diameter main tube. Trailing-link front suspension eased the ride, and would become characteristic of almost all subsequent Lambrettas.

► 1950 Lambretta Model LC
Even with its extra seat and more-powerful engine, the Lambretta trailed the Vespa in popularity—and for one obvious reason. Corradino d'Ascanio's bodywork saved riders from road-spray and engine grime, and the Lambretta left them out in the open to fend for themselves. In 1950, Innocenti finally realized that bodywork was essential for a scooter, and gave the public a choice of the "undressed" C or its new "dressed" LC; the "L" stood for Lusso or Luxury, surely a misnomer in terms of scooters. Luxury in Lambretta terms was two sheet-metal side covers that attached to a central housing along with full legshields up front. It was such a simple solution that it's difficult to understand why it was not done sooner. And to further prove the point, with the LC of 1950-1951 and the subsequent LD of 1951-1958, Innocenti finally had a winning combination. Sales of the new Lusso models soared, and by 1956, Innocenti halted production of the *open* scooters. *Collezione Vittorio Tessera*

▲ 1959 Lambretta LD125 Series IV
The C was followed by the D in 1951, which again had a redesigned frame, front fork tubes that enclosed the front suspension springs, larger 4.00x8in tires, and a new rear suspension setup with the engine hung by pivoting links and suspended at the rear by a single damper. The D engine measured 52x58mm, fathering 4.8hp for a 45mph top speed. With the LD125, Innocenti had a run-away success. In 1954, Innocenti built its most powerful scooter to date, a 148cc based on a longer-stroke, 57x58mm D engine that created 6hp—and of course Piaggio quickly responded with its own 150cc model. The LD was refined and perfected, and with the full bodywork, provided everything the Vespa had from its start. The LD debuted in December 1951, and the LD150 followed in November 1954. Also in 1954, a high-class electric-start model was available. In late 1956, the LD125/57 and LD150/57 were launched as two-tone luxury models with a sleek cowling over the handlebars that incorporated the speedometer/-odometer and horn while leaving the headlamp on the front apron. The 6hp LD 150/57 came standard with such luxuries as a speedometer, pillion seat, white sidewall tires, and, according to *Popular Science,* a mechanical marvel that made Italy great: "a clock that works." The 150 needed a 12-volt battery to power all these appliances. Owner: Michael Dregni.

"Americans export Coca-Cola. Japanese export Sonys. Russians export Kalishnikovs. Italians export Vespas and Lambrettas."
—Proverb

▶ Sophisticated Lady and 1960 Lambretta Li125 Series II
In America, the land of land yachts, a jaunt to Food Fair on a Lambretta was the ultimate in Betty Crocker cool for 1960, especially when accessorized with white-frame sunglasses and high heels. Innocenti debuted its new Li line in 1958 based on a new engine with a single horizontal cylinder and enclosed duplex chain drive to the rear wheel via a four-speed gearbox. The 148cc measured 57x58mm and created 6.5hp; the 123cc was based on 52x58mm for 5.2hp. The Li Series I scooters were clothed by a curvaceous body; in 1959 the Series II placed the headlamp in a cowling on the handlebars. The Li was a phenomenal success for Innocenti. It was modern, powerful, and reliable, and the Li line would dictate the direction of Lambretta scooters until 1971 when Innocenti halted production. After that, Serveta in Spain and Scooters India would continue building Li-based models. *Innocenti*

▲ Truckload of Lambrettas
By 1950, Innocenti was building up to 100 Model B scooters each day. But whereas the 100,000th Vespa was built in 1950, only 44,000 Lambrettas had been built by that date. By 1952, when this truckload of Model D Lambrettas rolled out the gates of the Lambrate works, Innocenti had both its 125cc and a new 150cc model, as well as a new, full-bodywork scooter. By 1956, Innocenti was up to $43 million in annual Lambretta sales—far beyond Cushman's near $35 million but still trailing behind Piaggio. And by the mid-1950s, Innocenti was shipping 6,000 scooters annually to the United States as well as additional exports to a total of ninety-six other countries. *Herb Singe Collection*

The Second Italian Renaissance

Fabricato in Italia

The Vespa and Lambretta were not the first motorscooters after World War II—they were simply the best. Other Italian firms followed suit and produced legions of scooters that swarmed over the roads of Italy in the postwar days.

The late 1940s and early 1950s were boom years for motorscooter makers. Amongst the most prolific were the firms Iso, which debuted its first scooter in 1948 after years of building refrigerators, and Moto Parilla, which launched its scooter in 1952. Guzzi, MV Agusta, FB Mondial, Benelli, and others all followed the leaders to the marketplace.

▲ 1946 Gianca Nibbio
The Nibbio, or "Kite Bird of Prey" in English, was the first Italian scooter to hit the market following World War II, just months ahead of the Vespa and prototype scooters by Fiat and others. But the Nibbio beat them all. Built in Monza, it used a standard tube frame covered by sheet-steel bodywork. Suspension was advanced for a scooter: telescopic front forks and a rear swing arm with an avant garde monoshock mounted beneath the engine. That engine was a 98cc two-stroke of 48x54mm creating 2hp and cooled by flywheel fan. The multiplate clutch fed power to a two-speed gearbox, which was shifted by a heel-and-toe foot lever. Tires were 3.50x8in. In 1949, Nibbio construction was transferred to the San Cristoforo firm of Milan. Having a company named for San Cristoforo, the protector of travelers (who was later impeached by the Vatican after having done centuries of good work), must have been good advertising for a scooter. When the Nibbio first rolled off the new assembly lines it was substantially updated from the Gianca model. The bodywork was all new, as was the 125cc two-stroke engine. With 5hp at 4700rpm and a three-speed gearbox, the Nibbio was now good for 75km/h.

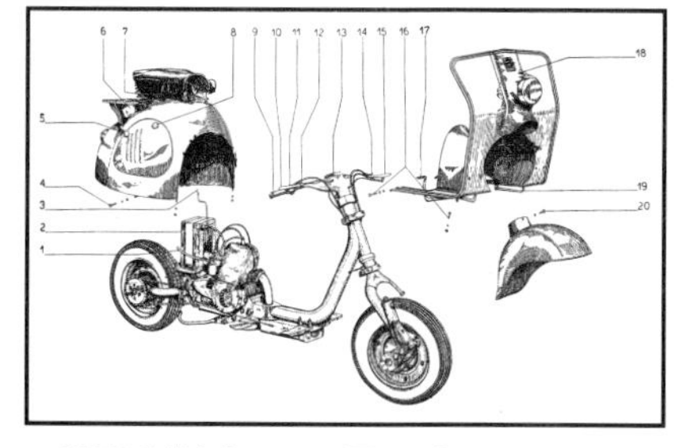

▲ 1951 MV Agusta Tipo C 125cc Super Lusso
The development of the MV Agusta firm followed a similar scenario to that of Piaggio—and the firms even named their first vehicles Vespa. In 1945, Enrico Piaggio was turning from wartime aviation production to a peacetime scooter called the Vespa; at the same time, Sicilian aristocrat Count Domenico Agusta was switching from aviation to a peacetime motorcycle called the Vespa. MV's Vespa was based on a 98cc two-stroke single, wet clutch, unit construction, and two-speed gearbox—the specifications could have easily been for the Piaggio Vespa. But Piaggio registered the name first, and Agusta's cycle went on the market as the MV 98. The two went their divergent ways from there: Piaggio to fame with the Vespa scooter; and MV to fame with its numerous racing motorcycles that would win thirty-seven manufacturer's world championships. Alongside its racing motorcycles, MV always produced street bikes as a way to finance Count Agusta's love of competition—much as Enzo Ferrari did with his GT cars. After developing a prototype Model A scooter, MV introduced its Model B 125cc scooter in 1949 with full monocoque bodywork and one-side front and rear lug axles—again, specifications shared with the Vespa scooter. In 1950, the Model C replaced the B with a tube-steel chassis covered by unstressed bodywork. In 1951, the CSL, or C Super Lusso, was offered.

▲ 1953 Parilla Levriere 150
Moto Parilla was created in the back of a truck diesel-injector repair shop on the outskirts of Milan in 1946. Parilla's Levriere or "Greyhound" scooter debuted in January 1952 as a refined scooter with a powerful 125cc engine, telescopic forks, 12in wheels on alloy rims, and stylish bodywork. In 1953, a 150cc version was added. The Levriere created a whole family tree of copies: Sweden's Husqvarna bought Parilla chassis in 1955 and mounted HVA engines; Germany's Victoria Peggy was a Levriere in disguise; and the Levriere was the major influence behind Zündapp's first Bella.

► 1950 Isoscooter and Lovers
Iso's history followed the ups and downs of the Italian postwar economy in textbook fashion. Iso of Bresso entered the scooter market in 1948 with a resume of building refrigerators. In 1950, Iso baptized the Isoscooter with a 125cc two-stroke split-single-cylinder engine. Iso built and sold large numbers of its Isoscooter to commuters hungry for economical transportation. After the success of the Isoscooter and along with the country's gradual recovery, Iso developed the Isetta minicar. Iso next stepped up to building *gran turismo* cars that challenged Ferrari. It was a success story that followed the Italian economy—and continued to follow it to the economy's collapse in the mid-1970s, when Iso too went out of business. *Collezione Vittorio Tessera*

► 1954-1960 Moto Rumi Formichino
There's a certain exquisite poetry in moving from manufacturing miniature submarines to building motorscooters. Like Enrico Piaggio, Donnino Rumi turned his swords into plowshares following the war, creating Moto Rumi in Bergamo in 1949 to build a series of eccentric motorcycles and scooters. Rumi's first scooter was the Scoiattolo, or Squirrel, of 1951-1957. Like the Vespa, the chassis and body were a stamped-steel unit without a frame. The Scoiattolo was followed by the Formichino, or Little Ant, of 1954-1960, which was, without doubt, a motorscooter masterpiece. The chassis and bodywork were of an innovative unit design akin to d'Ascanio's Vespa, but instead of the stamped steel of the Wasp, the Rumi body was cast from aluminum alloy in three structural sections that were assembled with Phillips-head studs. Both Rumi scooters were powered by the firm's unique 125cc horizontal two-cylinder two-stroke engine, which pushed the Ant to a top speed of 105km/h or 69mph. Normale, Lusso, Economico (1958), Sport (1957-1960), and racing Bol d'Or (1958-1959) versions were available. The Bol d'Or could be fitted with dual 22mm Dell'Ortos, blunderbuss-style megaphones, and alloy barrels and heads; top speed was up to a staggering 93.2mph.

La Bella Donna and Lo Scooter

Sex Appeal for Both Sexes

▲ 1954 Vespa and New York Calendar Girl
The 1954 Piaggio calendar took *Vespisti* on a tour of the world, from Rome to Egypt to New York, showing the "natives" and the uses they put to their Vespas. Christmas shopping in Gotham was made easy by the Vespa 125cc, but the ride must have been a bit chilly in such a short skirt. The mink stole no doubt saved the day.

Sex sold scooters just as well as sex sells most everything from laundry detergent to beer. But with motorscooters in the 1950s, advertising the putt-putts by draping a member of the female persuasion over the flowing bodywork served as a twisted double entendre.

Sure, a calendar pinup of Gina Lollabrigida straddling a Vespa inspired many an Italian boy with dreams of that 1954 150cc with its new 57x57mm square engine. But the scooter companies' advertising was sex appeal in another sense of the word as well.

The calendars and ads also promoted the idea of owning a Vespa to women. The scooter was easy to operate, didn't spray road grime or engine oil on you due to its stylish all-encompassing bodywork, and you could wear a dress while driving it, versus a motorcycle or even a car, which made you hike up your skirt on entry and exit. The scooter was a fashion accessory along with lipstick, painted fingernails, and high heels—and you could operate it while wearing lipstick, painted fingernails, and high heels.

> "As the family's second 'car,' the scooter makes shopping a pleasure for the housewife."
> —*Popular Mechanics*, 1947

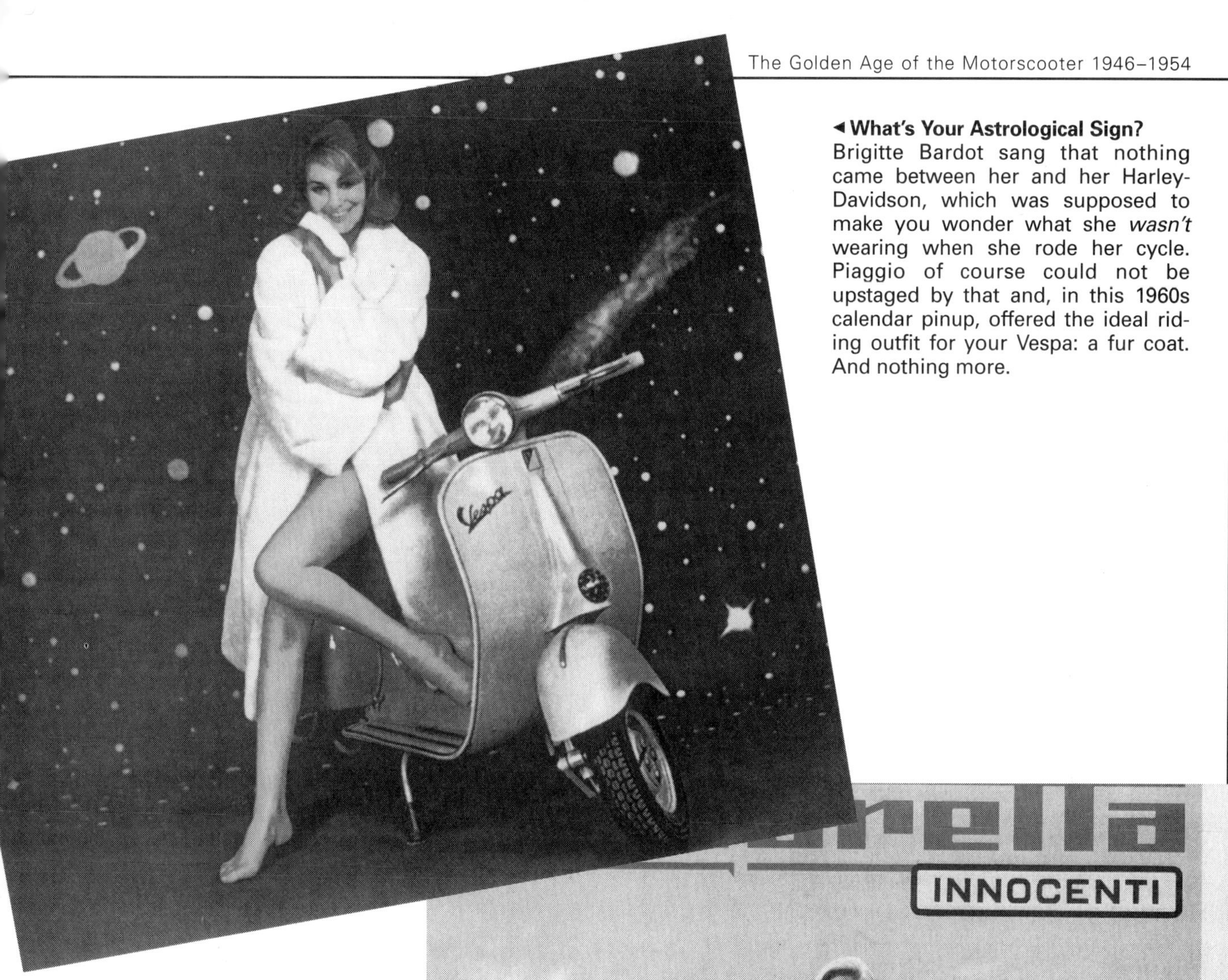

◄ What's Your Astrological Sign?
Brigitte Bardot sang that nothing came between her and her Harley-Davidson, which was supposed to make you wonder what she *wasn't* wearing when she rode her cycle. Piaggio of course could not be upstaged by that and, in this 1960s calendar pinup, offered the ideal riding outfit for your Vespa: a fur coat. And nothing more.

▲ Eve, the Apple, and a Dürkopp Diana
Tempting indeed, with its 194cc 12hp engine that was good for an alluring 100km/h top speed. The Diana's introduction in 1954 was splashed across Dürkopp ads when Miss Germany "won" a Diana and posed on her beloved scooter for a photo op.

▲ Jayne Mansfield and Lambretta Li125
Gentlemen prefer Lambrettas. *David Gaylin/Motor Cycle Days*

Mobilizing Japan

The American Scooter's Rising Son

Italy's Vespa and Lambretta have always overshadowed the development of other motorscooters, but following World War II Japanese companies responded as quickly as the Italians to their country's need for cheap transportation by building scooters. The first Japanese scooter was offered in 1946, the same year as the Vespa. The premier scooter, named the Rabbit, was built by Fuji of Tokyo. Mitsubishi, a builder of the famed Zero-San fighter for the Japanese Air Force, followed in 1948 with its Pigeon scooter.

Not surprisingly, considering the proximity to the US and their lightning-fast inception, these first Japanese scooters copied the style of prewar US scooters. Fuji scooter engineers had done their homework well, and the early Rabbits followed the tenets of scooter design laid down by the Salsbury way back in 1936–1937.

While these first scooters were clear copies of postwar US designs and were rustic at best, Japan soon developed efficient luxury scooters as good as any ever built by the Germans. The Japanese appetite for scooters was as strong as the Italians. In 1946, just eight scooters were built in Japan, all by Fuji; by 1954, more than 450,000 scooters were on the road with the country's total production at 50,000 annually. By 1958, total annual production was 113,218.

▲ 1946 Fuji Rabbit and Smiling Woman
In 1946, Fuji Heavy Industries of Tokyo created Japan's first motorscooter, the Rabbit, the same year as the first Vespas went into production on the other side of the world. In that year, only eight Rabbits were built, but in the following years the assembly lines multiplied production to make the scoot's namesake proud. The Rabbit was a rustic little scooter, similar in style to prewar American models; Fuji had done its homework. A padded cushion provided seating and suspension; a luggage rack was mounted behind. The Rabbit was based on a 135cc four-stroke single of 57x55mm producing 2hp at 3000rpm. Top speed was an ambitious 55km/h.

► 1950 Mitsubishi Pigeon C-21
During World War II, Mitsubishi was one of several builders of the Zero-San for the Japanese air force. Like Piaggio, the postwar Mitsubishi turned its talents in aircraft structural engineering to the field of motorscooters. In 1948, Mitsubishi introduced its first Pigeon, the C-11 with a 1.5hp 115cc two-stroke engine and pure Salsbury Motor Glide styling. In the 1950s, the Pigeons began breeding offspring models. The C-21 was introduced in 1950 with a 150cc two-stroke engine of 57x58mm for 3hp at 3800rpm. It was available with a passenger or delivery sidecar, following the line-ups of the American scooter manufacturers once again. The Pigeon emblem on the front fork apron was a Nazi-esque symbol that would continue to be used into the 1950s. Overall styling was agricultural, shaped more by a blacksmith's hammer than a stylist's hand, a "look" the line would keep until the mid-1950s. From the late 1940s and into the 1960s, Fuji and Mitsubishi would control the Japanese scooter market as a Far Eastern Piaggio–Innocenti zaibatsu duo.

◄ 1955 Honda KB
Soichiro Honda's story is a classic Horatio Alger rags-to-riches tale—translated into Japanese. Born the son of a village blacksmith, he got his start producing a radical new piston ring pre-World War II. During the war, his fledgling company built wooden propellers for the Japanese air force. After the war, Honda turned his hand to designing economical motorbikes and scooters. Honda first created a motorized bicycle, the Model A, in 1947 and built it in Honda's 12x18ft shed that housed thirteen employees. In 1953, the motorbike gave way to the first Cub clip-on motor, which developed into the Cub moped line that continued all the way into the 1980s as the Passport. By the end of its run, some 15 million Cubs had been built. Honda built its first scooter in 1954, the Juno KA, based on a 189cc overhead-valve engine with rear drive by enclosed chain. In 1955, the KB replaced the KA with a 220cc ohv engine of 70x57mm for 9hp at 5500rpm riding on 5.00x9in tires. The styling of these first Honda scooters was either bizarre or futuristic, depending on your point of view. It continued the Japanese scooter design philosophy of more is better and was dressed up with miles of molding running from the front turn signals to the rear turn signals.

▲ 1954 Fuji Rabbit S-61 III
The S-61 was the luxury flagship scooter of the Rabbit line. The silky smooth side-valve 225cc engine produced 5.9hp at a sedate 4000rpm and was started by the turn of a key and an "electric leg." Top speed was an awe-inspiring 75km/h atop those diminutive 4.00x8in wheels. The styling was cutting-edge Japanese design of the time, based on a philosophy of more is better. Thus, the Rabbit's flowing front apron was set against slab-faced legshields; the boxy tail was softened by adding multiple curved bodylines, contours, and air inlets. And to make sure it all caught your eye, the whole affair was bejeweled by miles of trimwork and the debut of the circular chrome Rabbit logo.

The English Scooter World

A Vacation Inspiration

▶ 1951 Douglas Vespa
The most famous and prolific English scooter was actually Italian. The venerable English Douglas firm first showed an imported Piaggio Vespa with the Douglas Vespa nameplate on the front legshield at the 1949 Earls Court Show in London, and plans were announced to build 10,000 Vespas in England. Douglas launched its Vespa line on March 15, 1951, building scooters under Piaggio license at the Douglas factory in Bristol. Douglas termed its model the 2L2, and moved the headlamp onto the front apron just below the handlebars to meet British law. Beyond the headlamp, Douglas followed Piaggio's lead even down to the same metallic industrial green paint. But as these Vespas were made in Britain, many of the components were sourced in the UK, such as the Amal carburetors, Lucas electrics, Milverton saddles, British-made Michelin tires, and BTH magnetos. In 1955, Douglas also offered a new model, the Piaggio Vespa 150 GS, which Douglas named the VS1 for Vespa Sports. While Douglas was building the 125cc Vespas, it only imported and never built the Piaggio GS, rebadging it as a Douglas model. British production of Vespas ended in 1963–1964, after which Douglas continued to import Piaggio models. Douglas had built a total of 126,230 Vespas.

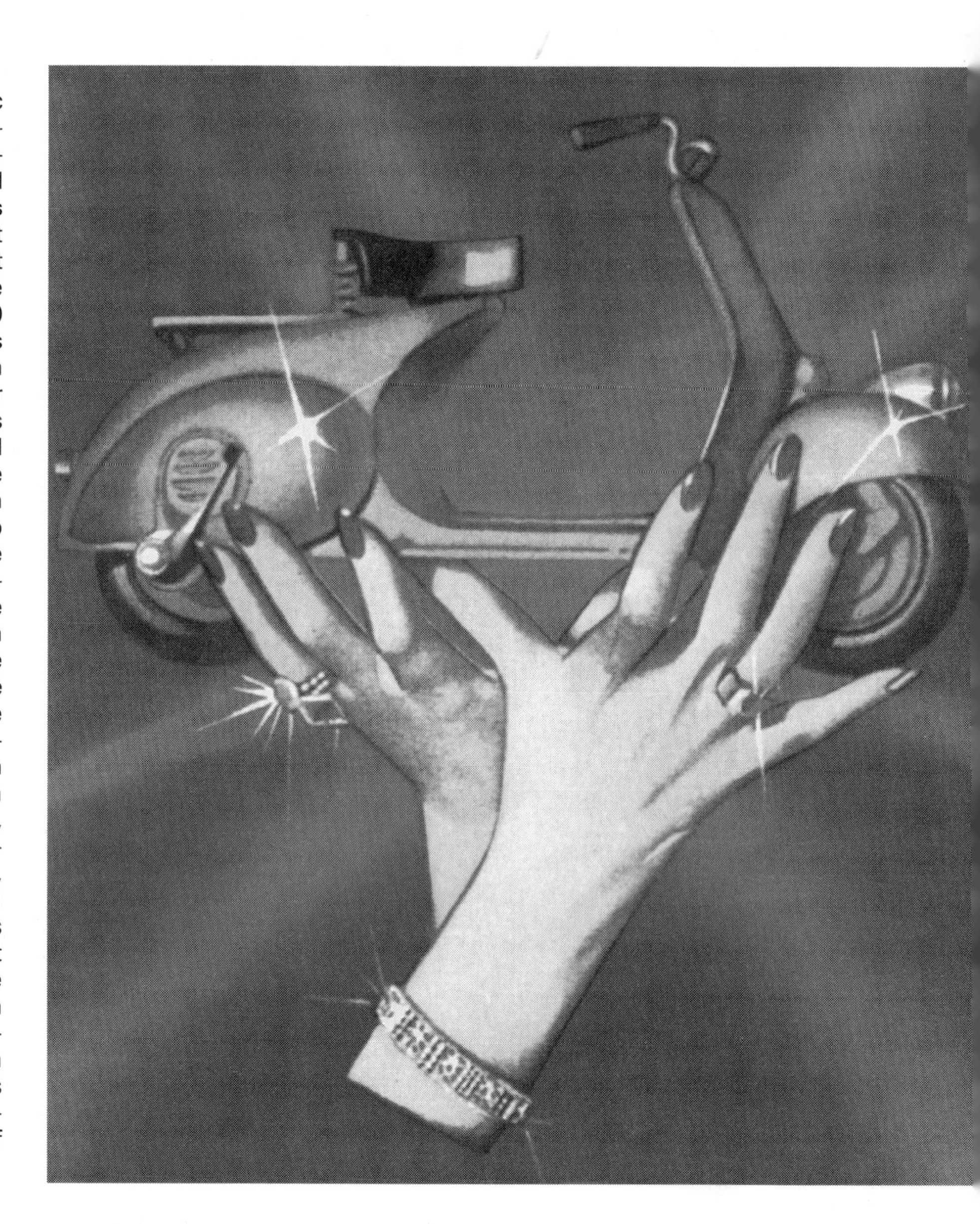

"These are to Certify that by direction of His Royal Highness The Prince Philip, Duke of Edinburgh, I have appointed Douglas (Kingswood) Ltd. into the place and quality of Suppliers of Vespa Scooters to His Royal Highness. . . ."
—1967 Royal Warrant for the supply of Vespas to HRH Prince Philip

While on vacation to sunny Italy in 1948, Englishman Claude McCormack, chief of the venerable Douglas motorcycle firm, got religion. While other tourists were swatting at the menace of the Italian scooters buzzing like insects around the Trevi fountain, McCormack saw a gold mine on wheels. Douglas began importing Vespas, showing its first "Douglas" Vespa at the 1949 Earls Court Show in London, and plans were announced to build 10,000 Vespas in England. It was the beginning of a long, prosperous history of English-built scooters.

Other English motorcycle makers such as Triumph, BSA, Velocette, Excelsior, and even the high-and-mighty Vincent with its Piatti tie-in launched scooters in the coming years. And a host of upstart makers also hawked their wares, including the infamous Lawrence Bond, whose various two- and three-wheeled scooter and microcar creations are recalled by the British today with much the same fondness the country feels for the V-1 rocket.

▲ 1959-1965 Triumph Tigress
Hailing from the beloved Triumph motorcycle company, the Tigress bore a name that was the feminine version of the famed Tiger motorcycle; the Tigress monicker could hardly have helped scoot sales. It's too bad, because underneath the bulbous bodywork was an engine heads above any other scooter of the era: a four-stroke 250cc vertical twin with overhead valves! Performance of the 250cc engine was all a scooterist could have wished for—in fact, the 250cc prototype was so fast it had to be detuned for production. But the Tigress was not a success, arriving in the market with too much too late.

▲ 1959-1964 BSA Sunbeam B2
Birmingham Small Arms was one of the stalwart British motorcycle pioneers, but in 1959 it created a motorscooter. BSA owned Triumph by this time, so to save money it created one scooter and offered it under both names. Neither worked. The scooters were designed by Edward Turner, who up to that time had a great reputation based on his Triumph Speed Twin vertical engine. In 1959, nearly identical Triumph Tigress and BSA Sunbeam models were offered, each in two versions, a 175cc two-stroke single and a 250cc four-stroke twin, the 250cc engines zooming the scooters faster than the 10in wheels and 5in drum brakes were made to handle. BSA-Triumph built its scooters in an attempt to thwart the Italian-dominated scooter market, but it was too late. By 1960, much of the craze had faded away, and the Tigress and Sunbeam died a quick death.

▲ 1959 Excelsior Monarch
Excelsior of Birmingham created the Welbike, a lightweight, collapsible scooter built for the British World War II paratroopers. Based on a 98cc Villiers horizontal two-stroke that fit snuggly within the center of the scooter, its top speed was a poky 30mph. According to English scooter historian Michael Webster, some 4,000 Welbikes were supplied to the British Army and used in Europe and the Orient as well as coming ashore at Normandy and braving the Western Front. After the war, the Welbike was civilized as the Corgi and produced by Brockhouse. Excelsior returned to the scooter market in 1959 with its stately, plump Monarch scooter, sharing body sheet metal with the DKR and Sun scooters but fitted with a 147cc two-stroke Excelsior engine. The KS version had a kickstarter and the EL an electric; in 1960, the versions were renamed the MK1 and ME1 respectively.

German Scooter Engineering

Volksrollers: The Tiny Wonders

> "In the shadow of the Wurmling Chapel, made famous by the well-known song by Ludwig Unland, lies an extensive factory, animated by the spirit of progress, and staffed by men who take an intense pride in their work. This is the birthplace of Maico."
> —1955 Maico Information Bulletin

Germany jumped onto the motorscooter boom at the end of the 1940s with Teutonic vigor. As German industry was wracked by the ruin caused by extensive Allied bombing, few factories had any capacity to produce their own scooters from scratch. Instead, firms licensed to assemble Italian scooters with their own nameplates: Hoffmann began assembling Vespas in 1949, NSU created Lambrettas starting in 1950, and Zündapp created its version of the Parilla Levriere starting in 1953.

By the early 1950s, the German makers had forsaken the Italian engineering for not living up to their standards and created their own scooters from the wheels up. Started by the push of an electric switch, propelled by powerful engines, and riding on supple suspensions, the new breed of German scooters were among the most luxurious *volksrollers* in the world.

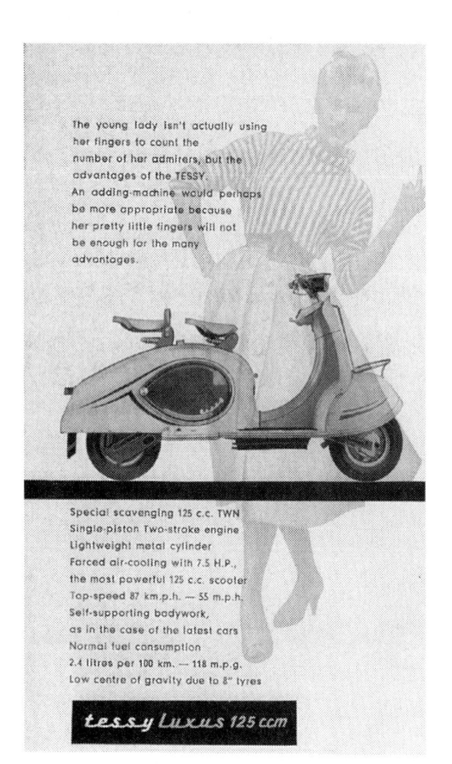

◄ **1956-1957 Triumph TWN Tessy Luxus 125**
Triumph of England was founded by two Germans in 1897, and when the firm started making motorcycles in 1903, they opened a factory in Nürnberg; in 1929, the company split into its British half, TEC Triumph, and the German side, TWN Triumph. In 1956, TWN brought out its first scooter, the Tessy with a 125cc, 7.5hp two-stroke single-cylinder engine. As this early flyer points out, "The young lady isn't actually using her fingers to count the number of her admirers, but the advantages of the TESSY. An adding-machine [which TWN also built] would perhaps be more appropriate because her pretty little fingers will not be enough for the many advantages." Quaint—or smarmy, depending on your point of view. *W. Conway Link/Deutsches Motorrad Registry*

▲ 1954 Progress Strolch
Creating a romantic image of *wanderlust*, Progress named its scooter the Strolch, or Vagabond. Three models were offered, all with engines by the prolific Sachs firm with 6.5hp for the 147cc model, 9hp for the 174cc, and 10hp for the 1955 191cc Progress 200. Early Strolchs had an impressive feature with their headlamps mounted on the front apron but moving separately, following the direction of the front wheel. *W. Conway Link/Deutsches Motorrad Registry*

> "Live joyfully with wings—drive an Adler."
> —1955 Adler Junior ad

◄ 1960-1965 Heinkel Tourist and Lovers
During World War II, Heinkel built many of the best bombers to give wings to Nazi Germany's Luftwaffe, but postwar Heinkel turned to two- and three-wheeled scooters and micro-cars as it was forbidden to produce airplanes. Heinkel's first scooter was the four-stroke Tourist 101 of 1953–1954 with 150cc, 7.2hp engine; it was followed by the 175cc, 9.2hp Tourist 102 and 103 of 1954-1965. For 1960-1965, this 150cc Heinkel Tourist debuted with modernistic Jet-Age styling, air vents, and two-tone pizzazz. The Tourist name was not mere hype; Heinkel gave silver plaques to be mounted on the front fender of scooters that surpassed 75,000km and gold plaques after 100,000km. The bronze plaque was discontinued as every self-respecting Heinkel owner should have been able to achieve 50,000km. *W. Conway Link/Deutsches Motorrad Registry*

▲ 1951–1958 Maico Mobil
On the cover of *Das Motorrad*, Maico called its Mobil, "Das Auto auf 2 Rädern," or "The Car on 2 Wheels." Dirigible on two wheels may have been more like it, but the enveloping bodywork was Maico's attempt to convince Germany in the days of the micro-car fad that two wheels were better than three or four. Maico debuted the Mobil at the Reulingen Show in June 1950 with a Herculean frame to support the armor-plated paneling. Underneath the bodywork, the Maico Mobil was more a motorcycle than scooter due to its 14in wheels, heavy-duty duplex-tube frame with a crossbar between the rider's legs, telescopic front forks, and its sheer bulk. The first Maico Mobil of 1951–1953 was powered by a 148cc single-cylinder engine fathering 6.5hp for the Fatherland via a three-speed gearbox, which was not quite enough power to push along its 253lb (115kg) mass. Maico's own brochures said it best: "With its latest product, the Maico-Mobil, Maico have introduced a completely novel type of machine which lies mid-way between the conventional motor cycle and the scooter; it may be that this will prove to be the true touring machine of the future." The future didn't last long for the Maico Mobil, but today the car on two wheels remains one of the most amazing scooters of all time. Be still beating heart. *W. Conway Link/Deutsches Motorrad Registry*

NSU's Five-Star Motorscooters

With Help from Lambretta and Volkswagen

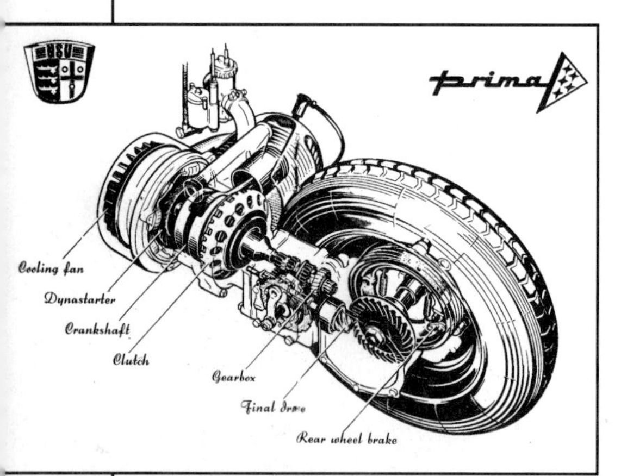

▲ 1957–1960 NSU Prima V Engine
German engineering at its best: NSU's amazing Prima engine. The one-cylinder two-stroke boasted a cylinder that was horizontal and transverse to the chassis. The flywheel magneto was at the front of the engine with the new four-speed gearbox to the rear of the crankshaft with a single-plate clutch mediating between the two. Final drive was via bevel gears. The complete engine unit was suspended from the frame by a front pivot mount and damped at the rear by a shock absorber with well-cushioned preload to handle the best German beer drinker. The 175cc electric-start Prima V engine had an oversquare design of 62x57.6mm and a four-speed gearbox. Power was now 9.5hp, all electrics were 12-volt, and top speed was 56mph.

In wartorn Germany, the quickest route to scooter production was licensing with an established maker. In 1950, NSU began fabricating Lambrettas in Neckarsulm under a five-year contract using Innocenti C/LC engines shipped from the Lambrate works and bodywork stamped out at the nearby Volkswagen factory.

But the Germans had no patience for certain features of the Lambretta, which they found lacking in good Teutonic over-engineering. Soon, NSU was building the majority of components for its scooters. In 1955, the NSU-Innocenti contract expired and was not renewed. Instead, in 1956, NSU introduced a Lambretta that was not a Lambretta; this was NSU's version of what a Lambretta should be, the Prima, Italian for First.

► 1950s NSU Prima Production
The NSU-Innocenti contract expired in 1955 and was not renewed. Instead, in 1956, NSU introduced a Lambretta that was not a Lambretta; this was NSU's version of what a Lambretta should be. Always competitive and class-conscious, NSU chose to call its new scooter the Prima, Italian for First. The engine was NSU's version of the shaft-drive Lambretta LC, still producing 6.2hp. The Prima D was only ever planned as a stopgap scooter to serve NSU's faithful during the year when the contract with Lambretta was scrapped and a new NSU scooter could be made ready. This new Prima debuted in 1957 as the Fünfstern, followed in 1958 by the Prima III.

▸ 1958–1960 NSU Prima III
By 1957, the Prima was refined as the high-class Prima 175cc Fünfstern (Five Star), joined by the 150cc Prima III, a simplified, democratic version of the V, of 1958–1960. The Prima was equipped with everything from electric starter to an electric fuel gauge, foglight, and radio. The Prima III was available as a K (kickstart) model from 1958–1960 and the KL ("Luxus" with electric start) for 1959–1960. The Prima V was exported worldwide and imported into the United States by Butler & Smith, Inc., of New York City, which also imported the later Prima III as the Deluxe. Butler & Smith was a renowned promoter; it even got a Prima V on the "Price is Right" TV game show in June 1959. The correct, winning price was $555.

▴ 1950–1954 NSU-Lambretta 125
In 1950, NSU signed on the dotted line to build Lambrettas for Deutscheslland. But NSU had no patience with certain features of the Italian scooters, and soon NSU was building the majority of components for its scooters. The NSU models surpassed the originals in the quality of brakes, Bosch 6-volt horn, Magura seats, and higher-output, 30-watt flywheel magneto. The Germans also added a glovebox topped by a small dash featuring ignition switch, speedometer with odometer, choke, and a handy clock. In 1953, NSU unveiled a new 12-volt electric-start model—this Luxus-Lambretta also included parking lights! For 1954-1956, NSU introduced a 150cc version with 6.2hp at 5200rpm. Top speed was 81km/h but the scooter retained the three-speed gearbox when a four-speed would have done it well. By the end of its life, production of the NSU Lambretta in all versions had proven prolific—117,045 units.

Zündapp's Beautiful Bella

The Mercedes-Benz of Motorscooters

In 1951, Zündapp strove to jump on the scooter bandwagon. The firm developed a scooter prototype itself, but decided to go the easy route that NSU had gone in licensing to build Lambrettas. Seeing that Vespa and Lambretta were already being sufficiently copied, Zündapp built its Bella scooter based on the 1952 125cc Parilla Levriere.

Throughout the 1950s, Zündapp continued to refine its scooter into one of the best machines on the *autobahn*.

▲ 1964-1984 Zündapp R50
Basking in the Greek sun, this 50cc model replaced the grand old Bella in 1964 as the times were a changing. And in creating this new model, Zündapp turned away from the now-defunct Parilla model and back to Lambretta and its new Slimline design with easily removable side panels, headlamp on the handlebars, 10in wheels, and handlebar gear change. The 49cc engine was based on the Zündapp Falconette moped; the three-speed R50 created 2.9hp at 4900rpm whereas the four-speed RS50 Super made 4.6hp at 7000rpm. Unfortunately, the company fell on hard economic times in 1985 after producing a total of 130,680 scooters; the entire Zündapp stock and machinery were sold to China. Approximately 1,500 Chinese workers went overland to Munich by train to pack up all the equipment to ship to the People's Republic. During the two weeks of loading, the Chinese workers slept in the packing crates to save money.

"It would be difficult to speak too highly of the Zündapp's navigational properties."
—*Motor Cycling*, December 1953

▲ *Bella Donna* and Zündapp Bella
Seeking to link its Germanic scooter to the success of the Italian machines, Zündapp named its beauty the Bella, Italian for Beautiful. The Bella was a hit from its debut at the 1953 Frankfurt show. The 147cc model created 7.3hp at 4700rpm for a maximum speed of 50mph. Until its demise in the early 1960s, the Zündapp Bella was among the most refined motorscooters anywhere in the world, a two-wheeled Mercedes-Benz for the well-heeled, plutocratic scooterist. *Collezione Vittorio Tessera*

▲ Side-Saddle on a Zündapp Bella
The first Bella of 1953-1955 was the R150, soon followed by the R200 of 1954-1955. For 1953 only, Zündapp created the Suburbanette, designed and named for the great American suburbs; sold only in the U.S., it was never a success story and, today, remains the rarest Zündapp scooter. By the early 1960s, sales had plummeted and only about 2,000 of each of these models were made. The Bella's beautiful life was over. *Collezione Vittorio Tessera*

► High Speed at a Standstill
Despite the image of speed and wind in their hair while riding their Zündapp Bella 200cc, these two scooterists actually had a rock placed behind the rear wheel so they didn't roll backward while the photo was taken. *Collezione Vittorio Tessera*

The "Social Appliance"

Rebuilding the World in Their Own Likeness

▲ Cruising Main Street on the Grocery Getter
The ideal grocery getter—with an air of elegance. In the post-World War II years, Mrs. Mary R. Spoor, 71, tooled around Danville, Illinois, on her Cushman Glide-Kar three-wheeler to do her shopping and visiting. *Herb Singe Collection*

Innocenti called its scooter a "social appliance" in what was probably an off-target Italian-to-English translation. Nevertheless, the term carried a much larger truth.

Bad times begat motorscooters, and spawned by the thousands in struggling, die-hard factories throughout Europe, scooters provided the wheels on which Europe's immediate postwar reconstruction rode. Scooters were inexpensive social appliances in many senses of the word—inexpensive to purchase, repair, and operate—and soon became a necessity. Scooters carried food to market, brought the country to the city and the city to the country, spread ideas and culture, transported Romeo in search of Juliet, and almost overnight carried Italy and much of the rest of Europe into the modern world; it was the dawn of a second European renaissance.

By the time people had more money in their pockets—due in part to their trusty steed, the scooter—those same scooterists turned their backs on their Vespas and made a down payment on a Fiat 500, Volkswagen, or Citroën 2CV. Many people turned their scooters over to their children, or junked them by the thousands to scrap metal yards, remembering their faithful scooter with all the fondness of a carpet bombing.

The road to recovery had been steep, but the two-stroke, exhaust-burping, oil-burning scooter had climbed the hill.

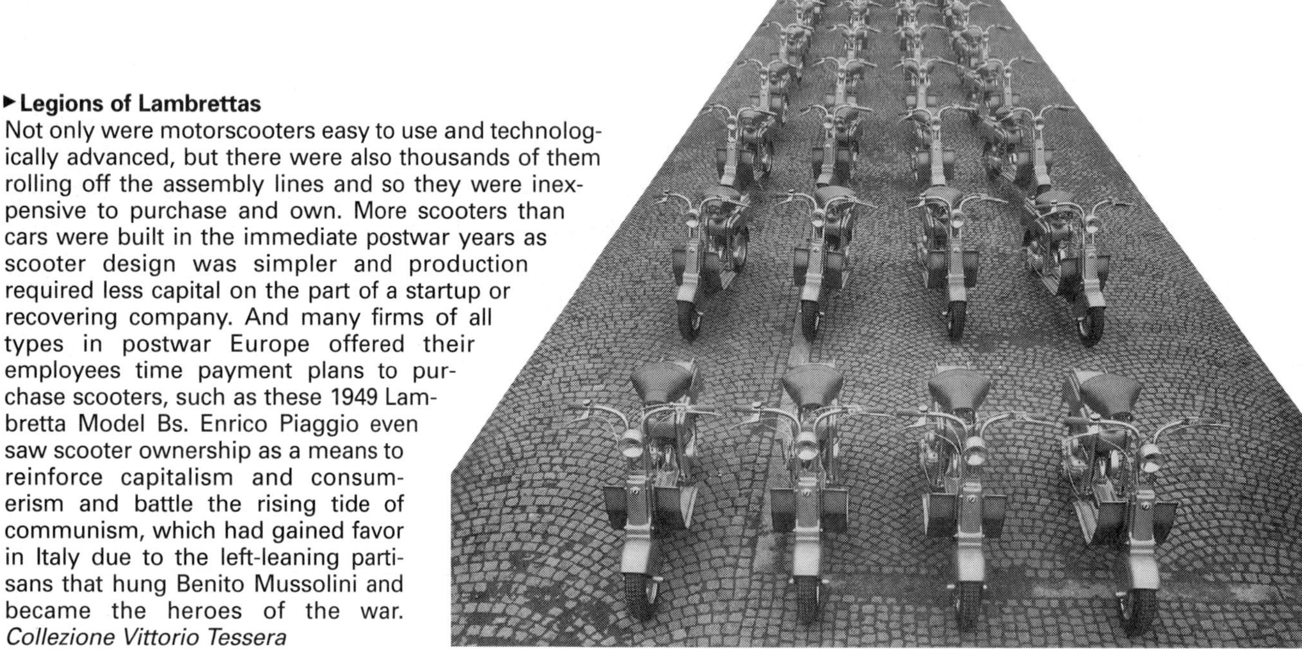

► Legions of Lambrettas
Not only were motorscooters easy to use and technologically advanced, but there were also thousands of them rolling off the assembly lines and so they were inexpensive to purchase and own. More scooters than cars were built in the immediate postwar years as scooter design was simpler and production required less capital on the part of a startup or recovering company. And many firms of all types in postwar Europe offered their employees time payment plans to purchase scooters, such as these 1949 Lambretta Model Bs. Enrico Piaggio even saw scooter ownership as a means to reinforce capitalism and consumerism and battle the rising tide of communism, which had gained favor in Italy due to the left-leaning partisans that hung Benito Mussolini and became the heroes of the war. *Collezione Vittorio Tessera*

"The best way to fight Communism in this country [Italy] is to give each worker a scooter, so he will have his own transportation, have something valuable of his own, and have a stake in the principle of private property."
—Enrico Piaggio,
Time magazine, 1952

►NSU Prima V as a Second Car
The NSU Prima Fünfstern of 1957-1960 was the ideal second car for the German plutocrat. With engineering to inspire envy in Mercedes-Benz owners of the time, the Prima was perhaps the foremost luxury scooter of all time—and it boasted almost as many chrome doodads, extraneous lights, and jet-styled airscoops as the best cars of the late 1950s. The suspension was even tuned for heavy-duty *doppelbock* consumption. *W. Conway Link/Deutsches Motorrad Registry*

▲1952 Norwegian Homebuilt Scooter
After the dark days of World War II, commodities such as food and motorscooters were not easy to come by. In Norway, taxi-driver Kristoffer Gjevre had seen a newspaper picture of an Italian scooter, but such a social appliance was not available in the far north. So, in his spare time, he built this "Bigge" in his garage using wheelbarrow wheels, DKW motorcycle brakes, sawn-off 1920s Harley-Davidson front forks, and a DKW 98cc engine. The front fender, legshield, and floorboards were made of home-bent sheet metal, whereas the engine cover was crafted from plywood. All in all, it was a modern motorscooter that balanced so well it could be driven hands off. Painted a brilliant green, it was a confirmation present for his son, Ole, who drove it every summer day—and still proudly pilots it today. *Ole Birger Gjevre*

Scooters Made in America

A Separate and Distinct Species

The United States was the Galápagos Islands of motorscooters. American scooters were a species that evolved on its own separate from the other continents—and without the benefit of "inspiration" from other scooter makers. While American scooters—especially the prewar Salsbury—were influential to many Japanese and a handful of European scooter makers, US makers largely ignored the styling and engineering advances made in Europe and the Far East—until it was too late.

Salsbury's scooters were so successful prewar that they were bought up by AVION, Inc., which soon came under control of the fledgling Northrop Aviation Corporation; but by 1949, Northrop had given up on scooters.

Other makers were more successful. Moto-Scoot and Mead continued to prosper, as did newcomers such as Mustang. In fact, scooters were on such a boom that mail-order merchandisers like Sears Roebuck and Co., Montgomery Ward, and Gambles began selling massive quantities of scooters by mail. It was the dawn of a new scooter Golden Age in the USA.

La falciatrice scooter Fairbanks del 1956

▲ 1956 Fairbanks-Morse Scooter
After building behemoth farm tractors and weight scales for decades, Fairbanks-Morse weighed into the scooter market in 1956 with this creation. Powered by a 2.75hp motor, which was good power for the day, the Fairbanks-Morse could only reach 5mph top speed, which was lousy in any day. With styling like a lawnmower, the firm missed the scooter mark and perhaps should have gone into the mower market.

◄ 1960-1965 Harley-Davidson Topper
The Harley-Davidson Topper was a Hell's Angels starter scooter, complete with the coveted pull-start engine that would wow them on main street Sturgis. The first Topper was released in 1960 to fanfare promoting it as "Tops in beauty and tops in performance." The Topper boasted a two-cycle 10ci engine laying on its side with the 2.38x2.28in single cylinder facing to the front, which was supposed to eliminate the need for a cooling fan. The Scootaway automatic transmission used a centrifugal clutch and V-belt with final drive by chain; ratios were "infinitely variable." As this 1960 ad puffed, "Mom's a Topper fan, too! She likes its good looks: sharp, clean lines molded in tough beautiful fiberglass." That's perhaps because mom recognized the Topper's styling influence: her new Frigidaire. By the end of 1965, the Topper was but a memory. Harley-Davidson chose to follow the road traveled by Cushman in canceling its scooter and shifting to golf carts. Harley-Davidson is probably still kicking its corporate self for building this one. *Harley-Davidson, Inc.*

▶ **1940s American Moto-Scoot**
Moto-Scoot was back after World War II as the American Moto-Scoot, yet little else had changed but the name. In 1946, Salsbury would debut its radically innovative Super-Scooter Model 85, and even Cushman's new 50 Series "Turtlebacks" had developed since prewar pioneer days. Moto-Scoot, however, was a holdover to the past. The styling had changed to look like a humpbacked wedding cake on wheels, but even this facelift when meshed with the old-fashioned motor dated the Moto-Scoot. Still, the company kept churning out scooters in a vast array of models. But the honeymoon was over, and Moto-Scoot was gone by the mid-1950s. *Herb Singe Collection*

"Advice to Teenagers: When it comes to combs and scooters, never a borrower or a lender be. Instead, start dropping hints to Mom and Dad about the new Topper."
—Harley-Davidson Topper ad

▲ **1949 Lowther Scooter**
This 1949 Lowther flyer modestly hailed its wares as "The 'Park Avenue' Scooter with the 'Main Street' Price." The styling was enough to make any scooterist green with envy: the chrome doodads, three-tiered bumper, and submarine-periscope headlamp were too cool. Lowther built scooters for several years, even badge-engineering putt-putts for the great Indian motorcycle firm, but when's the last time you saw one? Those Park Avenue types probably have them all hoarded away. *Herb Singe Collection*

▶ **1947-1950 Mustang Model 2**
The Mustang was certainly a different breed of steed. The story began with a budding young California engineer named Howard Forrest who loved to race anything with an engine. In the 1930s, he ran midget sprint cars, which formed the source of two mainstays of Mustang design: powerful small-bore engines and small wheels. He toyed with motorcycles before joining forces with entrepreneur John Gladden to produce the first Mustang Colt of 1945–1947. Half motorcycle, half scooter, it ran on 12in wheels, was powered by a prewar English Villiers single, and was named in honor of the P-51 Mustang fighter of World War II. In 1947, the Model 2 replaced the Colt with a new engine, quaintly named the Bumble Bee, and the first use of telescopic forks on an American motorcycle. The Model 2 was updated for 1950-1958 as the Model 4 and offered in a Standard and Special model. Mustang continued to roll out new models, including the revived Colt of 1956–1958, Pony of 1959–1965, Bronco of 1959–1965, Stallion of 1959–1965, Thoroughbred of 1960–1965, and Trail Machine of 1962–1965. Mustang's demise came with the arrival of Honda; the year 1965 was a sad one for American motorscooters as Cushman curtailed production of its Silver Eagle line and Mustang production was grinding to a halt. In 1966, production ceased.

Salsbury's Super-Scooter Spaceship

Buck Rogers' Wildest Dream Come True

▲ 1946-1949 Salsbury Super-Scooters Model 85 Deluxe and Standard
It was outrageous. It was radical. It was the motorscooter of the future—and destined to be the most desirable motorscooter of all time. Here are four 1940s swingers and the hottest wheels available anywhere—the Model 85 Standard, right, and DeLuxe, left, both using the same mechanicals but the latter with a plexiglass windshield and fully streamlined front cockpit. Northrop produced the Model 85 for two years, during which time Foster Salsbury estimated that a mere 700–1,000 units were built and sold, including exports to Germany. But as Salsbury remembered of the Model 85's demise, "Demand fell off when cars started becoming available again, and Northrop halted production." It was a sad finale for a great scooter. *E. Foster Salsbury Archives*

It was the motorscooter to make Buck Rogers green with envy, an unblinking vision of a bountiful future for the scooterist, as new as tomorrow and twice as flashy. It was the Salsbury Super-Scooter Model 85, in Standard and Deluxe versions.

From stem to stern it was chock full of technical wonders, innovative inventions, and the truly bizarre. It was the motorscooter that made Cushman's "Turtleback" look like a Model T, the Vespa look like it was designed by Rip Van Winkle, and the Lambretta as if it had crawled out from under a prehistoric rock.

It was also too much of a good thing, didn't sell, and was gone from the market within three years.

"Even the fast-flying airplane has to rely on the lowly scooter when taxiing into a parking area at the airport."
—*Popular Mechanics*, 1947

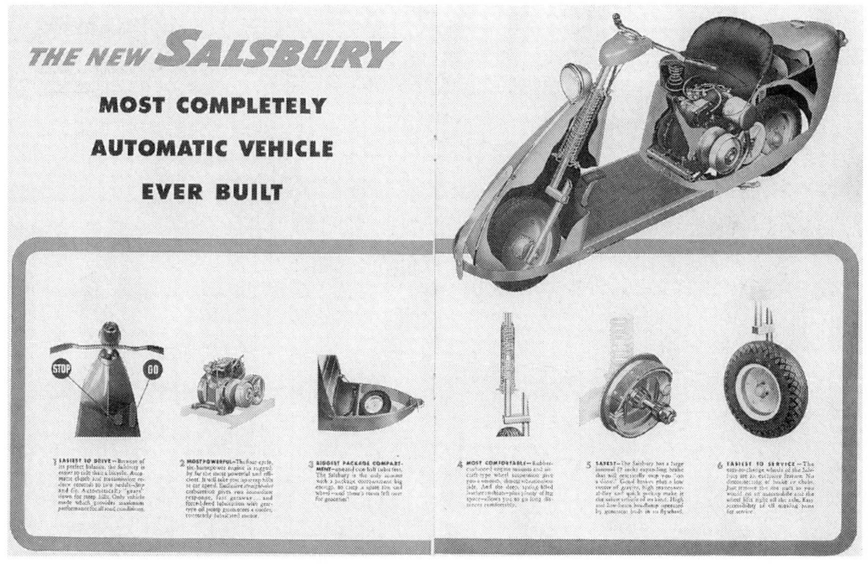

▲ 1947 Salsbury Model 85 Brochure
The new Super-Scooter was hailed by Northrop as the "Most Completely Automatic Vehicle Ever Built." The engine was a Salsbury-built 6hp four-stroke single-cylinder set at an angle toward the rear of the scooter. It featured the exclusive Straight-Shot carburetion system with a short, but admittedly angled intake manifold. Nevertheless, the 1946 brochure promised the power "will take you up steep hills at car speed." The torque converter was standard on both models, and for further ease of operation, the 1946 brochure didn't even use the words throttle or brake when describing the two control pedals: they were simply the Stop and Go pedals. The 85 DeLuxe, however, could be ordered with an auxiliary hand throttle. Perhaps the most interesting feature of the new 85 was the one-sided front forks with a stub axle inspired by airplane strut design. But the suspension was provided by two, different-rate coil springs housed *within* an elongated steering head. At the rear, a single coil spring did the damping. Within the tail bodywork was a spare tire and luggage compartment "ample for most shopping trips." The 85 had come a long way from the Motor Glide. *E. Foster Salsbury Archives*

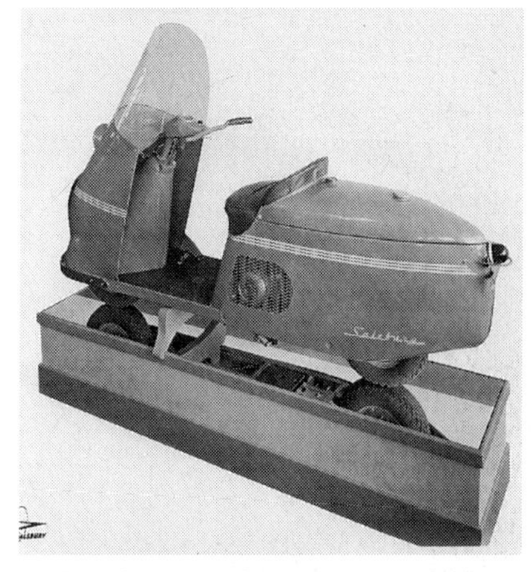

▲ 1947 Salsbury Super-Scooter Model 85 Deluxe
In 1944–1945, Salsbury sold his scooter business to the Los Angeles defense-contracting firm AVION, Inc., which in turn became a subsidiary of Northrop Aircraft Corporation. Based from a new plant in Pomona, the new Model 85 scooter had been designed by engineer Lewis Thostenson during the end of World War II and was introduced in late 1946. Thostenson had been one of Salsbury's original engineers and continued on at Northrop; Foster Salsbury now headed the scooter sales team. Northrop called its new Super-Scooter Model 85 the Salsbury scooter; not the Motor Glide by Salsbury. Obviously, Northrop believed Salsbury was an established name—and it didn't want further confusion with Cushman's Auto-Glide. *E. Foster Salsbury Archives*

◄ Policeman and Salsbury Model 85
Its styling was avant garde, years ahead of the Jet-Age styling of the 1950s. Salsbury spent the war years producing war materiel and had also developed an experimental wind tunnel for the aviation industry, which was used in early aerodynamic studies for fighters and bombers. Lessons learned in the new field of aviation aerodynamics would set the stage for the design of the new Salsbury scooter. The Model 85 was also large. The size was unprecedented among scooters—it was certainly big enough to be a second car! *E. Foster Salsbury Archives*

The Classic Cushman "Turtleback"

The All-American Motorscooter

World War II was good to the Cushman Motor Works. Cushman came out of the dark war days on top of the US scooter world after years of building military motorscooters.

In 1948, Cushman launched its new line, the 50 Series, which won the hearts of American scooterists and was bestowed with the affectionate nickname of the "Turtleback" due to its streamlined rear-engine cover design. Cushman kept the look through the 60 Series.

But the end was in sight for the good old Cushman. In 1949, Cushman debuted its new miniature motorcycle, the Eagle, which soon took over the sales lead from the venerable "Turtleback."

In 1957, two important events took place for Cushman. On May 10, 1957, the Ammon family sold Cushman to Outboard Marine Corporation (OMC), the long-time maker of Johnson and Evinrude outboard motors. Also in 1957, the step-through scooter was redesigned as the 720 Series with "modern" looks that resembled a refrigerator on two wheels. The classic Cushman "Turtleback" was gone—but never forgotten.

▲ 1957 Cushman 720 Series Step Thru
In 1957, Cushman debuted its radical new 720 Series of scooters to replace the classic "Turtleback" 60 Series. The new 722 Pacemaker and 725 RoadKing models carried over the 60 Series names but little else. Almost everything had been redesigned and modernized—especially the new bodywork, which was either as "modern-as-tomorrow" or just boxy, depending on how you looked at it. Cushman said the new design had "eye-catching beauty" and termed the styling "streamlined." Whether you agreed or not, the value of any streamlining was dubious anyway at a 40mph top speed.

◄ 1948 Cushman Model 52
Cushman emerged from World War II to offer the best scooter on the American market, the new 50 Series, nicknamed the "Turtleback." The 52 was the standard model alongside the 52A Sport version without the engine cover "tail sheet," similar to the bare-bones mil-spec 53A. In addition, there was the flagship 54 with an automatic clutch and two-speed gearbox. In late 1948, Cushman bowed its new belt-drive torque-converter automatic clutch and transmission, called the Variamatic and denoted as the B-suffix to the Model 54B. But the Variamatic proved troublesome and "almost ruined the Cushman reputation for reliability," according to Cushman historian Bill Somerville. Finding a 54B with the original Variamatic today is like searching for the proverbial needle in the haystack; most were retrofitted with the less-sophisticated but tried-and-true Floating Drive. Owner: Roger McLaren.

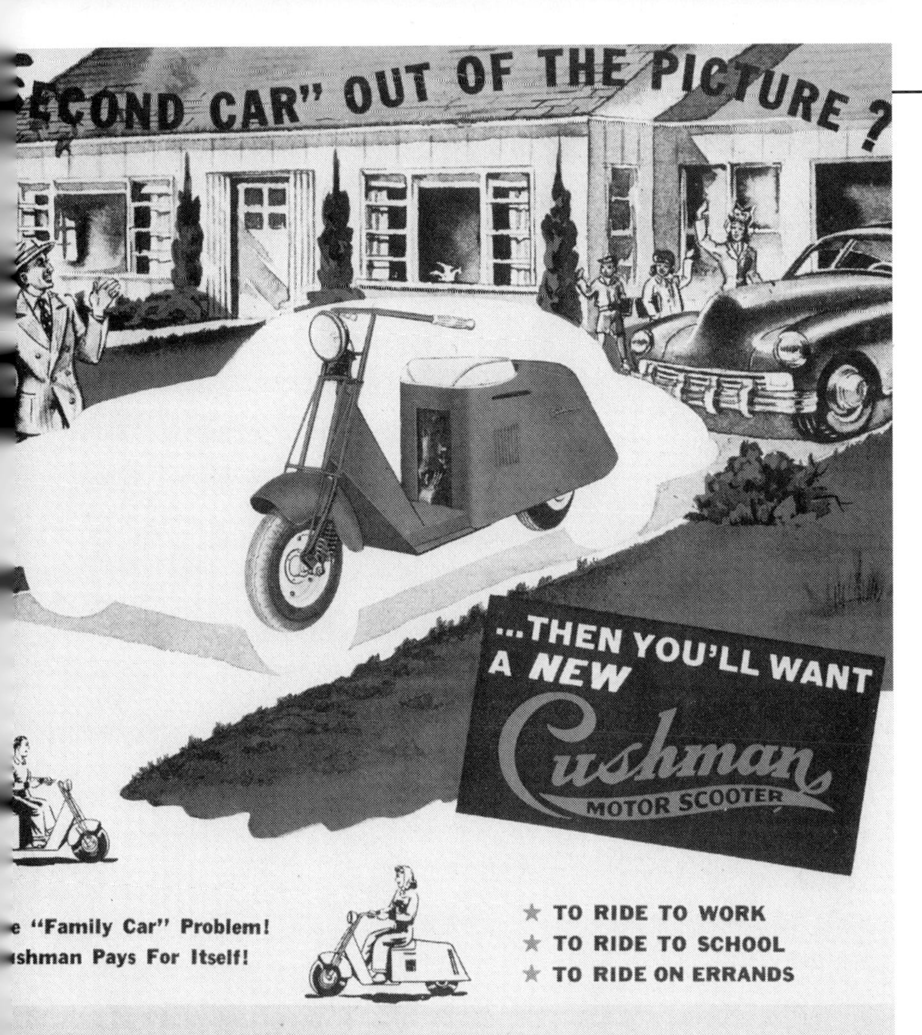

◄ **"Second Car" Out of the Picture?** Following the war, Cushman continued to promote the economy of the scooter as a second car, but the market in America was shifting. The Cushman may have been the land yacht of motorscooters, but in the boom years of the 1950s, most people could afford a second car, and scooters became a novelty item for hobbyists and teens. *Herb Singe Collection*

▲ **1951 Cushman 60 Series Ad** Cushman advertised the new 60 Series as "Twice the Scooter!" ushering in a new era of gigantic scooter sales numbers. Bowing in late 1948, the 62 Pacemaker replaced the 52 while the 62A was variously called a Sport or Super version of the Pacemaker but eschewing rear bodywork. The 62A was initially offered with only the Variamatic but soon switched to follow the rest of the 60 Series. Likewise, the 64 replaced the 54 but included a new two-speed sliding-gear transmission, and the 64A appeared in its Birthday Suit as a Sport. The 60 Series were powered by a variety of engines: the 4hp Husky was the staple engine; a 5hp version was optional.

"Cushman is ready for the boom in business. Its president, Robert H. Ammon, has had the company working on 'Detroit' principles for several years. Under its former job-shop setup, Cushman couldn't begin to turn out its present volume—over 10,000 scooters a year. But by scaling assembly-line techniques down to small-plant operation, Ammon has pushed dollar volume up to $35 million a year—90 percent of it from scooter sales—and steered the company into first place among U.S. scooter manufacturers."
—*Business Week*, August 19, 1950

Wanderlust, Scooter Style

The World is Your Oyster

While some viewed their scooter as means to get to their end destination, others thought of it as a way to see the world. With ads boasting 120mpg, travel cost a fraction of what it once did.

Many landlubbers had never seen the sea and some had never even set foot out of their province. Now, travelers trusted their two-strokes to carry them for weeks on end, and with service stations geared toward the new scooter market, repairs were painless. Finally, everyone could throw away those *National Geographics* and see the world for themselves.

▼Recreational Vespa
Vespas have a hard enough time carrying two people, let alone a mini-Winnebago. This trailer was shown at a Parisian Touring and Holiday Show in the Spring of 1955 back when scooterists still believed that scooters could do anything. *Herb Singe Archives*

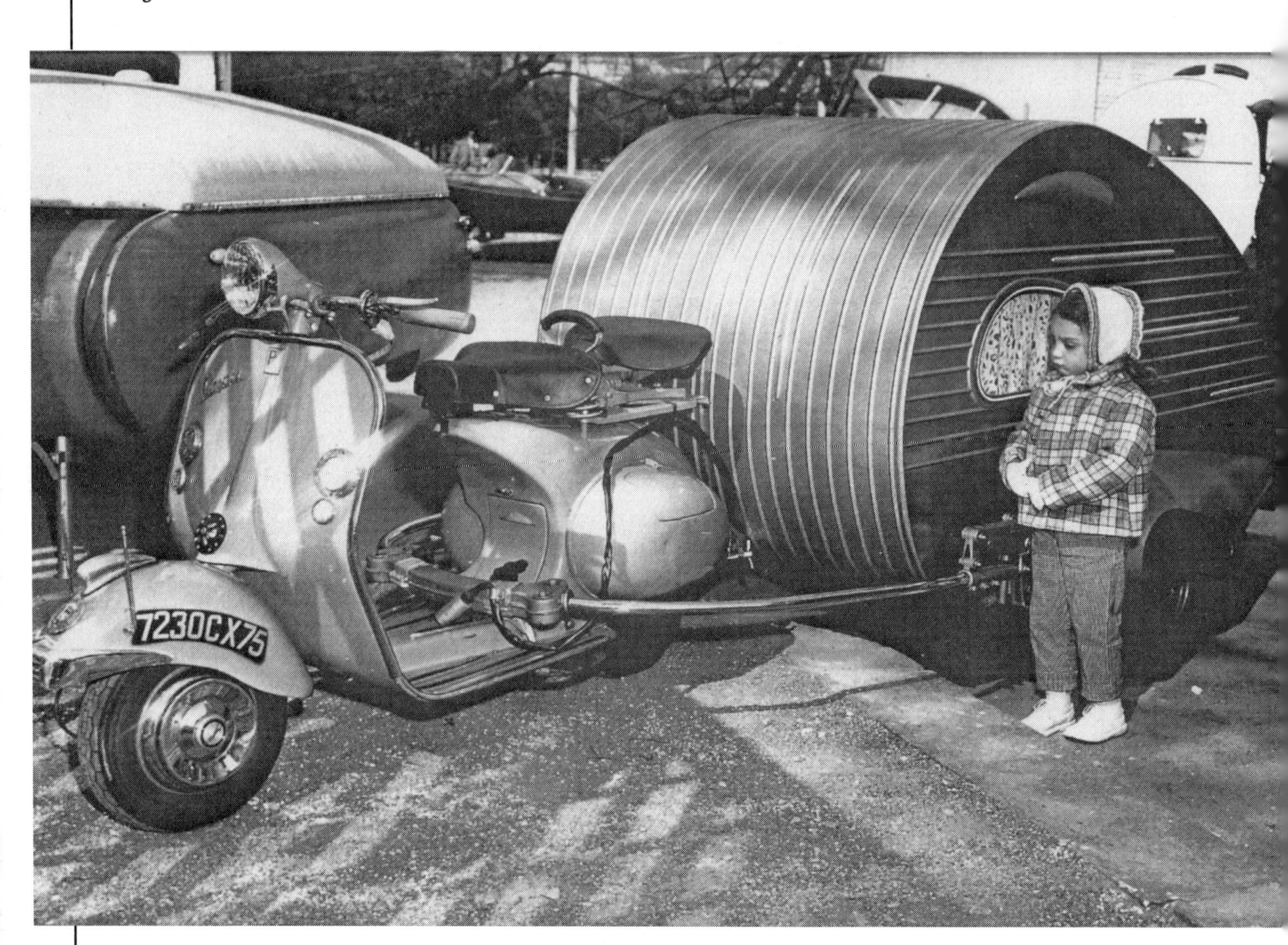

◄ Hold That Pose!
This European tourist whips off some snapshots with her Leica M3 *mit* Leicameter MC to bring home photographic evidence of the world one can see on a scooter. On her Dutch-built Bitri scooter, she would feel right at home in the Big Apple with policemen of the era protecting Gotham on their putt-putts. *W. Conway Link/Deutsches Motorrad Registry*

▲ Look Ma, No Brakes!
Although the craze for scooter vacations hit Europe the hardest, Cushman heralded their mobile as the one that followed in the pioneers' tire tracks (as long as the scooter could go downhill). Their motorscooters opened up a means for teenagers and other folks who couldn't afford a car to travel around their state and see the scenery.

▲ Arrgh!! Travel Poetry
Wanderlust mixes with jealousy of everyone's love and "enthusiasm for the new travel companion." German manufacturers knew that their market wanted to see the world and have the freedom not afforded by trains, thus scooters. This Adler Junior competed with the Zündapp Bella, the NSU Lambrettas, and the Heinkel Tourist for the vagabond market.

▲ Next Stop: Fun!
Scooters began to wean America from dependence on the railroad, but the nostalgia for trains remained. Although the ride on scooters is more bumpy, with gas mileage like that, who could resist an interstate adventure on 8in tires. Scootering is arguably one of the best ways to travel in the out-of-doors, but it's doubtful that a daily trip of 200 to 300 miles could be done "comfortably."

The World's Fastest Motorscooters

Land Speed Records on 10in Wheels

Innocenti started it all. In October 1950, scooter pilots Masetti, Romolo Ferri, and Dario Ambrosini ran a Lambretta Model B 1,592km at an average speed of 132.6km/h. Lambretta held the world's scooter speed record, and naturally this galled Piaggio. Early in 1951, Vespa struck back and beat the Lambretta record, upping the ante to 170km/h. In August 1951, Ferri set a score of world records—one of which was not broken for more than twenty years—on a super-streamlined Lambretta. His top speed with the 125cc torpedo was an astonishing 121mph. In 1965, Englishwoman Marlene Parker topped 130mph with a 200cc Lambretta, but the FIM ruled it unofficial.

Throughout the 1950s, Piaggio and Innocenti battled it out for top-speed supremacy in a never-ending attempt to best the other. Surprisingly, this war for speed and endurance was largely ignored by other European scooter makers who left Innocenti and Piaggio to their own devices. On the other side of the pond, however, American makers Mustang and Powell dueled with each other for the speed title on the dry lakes of California and the Bonneville Salt Flats.

◄ Timekeeper and 1954 Vespa
Keeping time for Scuderia Ferrari's Formula 1 racing team was this *bella donna* with her Vespa.

▲ 1949 Vespa Record-Setter
By the late 1940s, the war between Piaggio and Innocenti moved to the racetracks, where full-bore road-race and record-setting Vespas and Lambrettas outdid each other at record paces. One week, a Vespa owned the world scooter speed record, only to be beaten by a Lambretta the next. Throughout the 1940s and 1950s, both firms built a long line of works racers; this record-setter was thinly based on a production Vespa, and Piaggio even built a special prototype for a Sport scooter backed by a belief that as the economy recovered, there would be demand for a high-performance sporting model. The timing of the Sport prototype was too early, but the idea was sound; Piaggio's great Gran Sport model would use features of the racers and record-setters.

100mph Grocery Getters

In the 1990s, scooter racing continues in Italy, England, and the United States as well as other countries. The American Scooter Racing Association was formed in 1989 by Vince Mross of West Coast Lambretta Works in San Diego. ASRA holds several races annually in southern California with hot-rodded Vespas and Lambrettas in a range of stock and modified classes. Custom-framed Lambrettas with Jet 200 or 200 SX engines are at the top of the class.

◄ 100mph on 10in Wheels
A souped-up Lambretta leads two Vespas through a twisty in American Scooter Racing Association action. Any relation to a 500cc Grand Prix Honda—or a stock scooter—is purely accidental. *ASRA*

▲ 1949 Lambretta Road Racer and Model B
"33 World Records conquered by the motor scooter Lambretta" shouted this display. "Motor scooter" was used loosely here; compare the stock Model B in the foreground with the factory-built road racer—yes, the nuts and bolts were interchangeable between the stock and race versions, but little else. Road racing and record-setting were just two further arenas for the Vespa and Lambretta gladiators to do battle throughout the 1940s, 1950s, and 1960s. *Collezione Vittorio Tessera*

▲ 1959 Vespa GS Racers in the Giro di Tre Mare
Wherever there was a street that could be closed off on Sunday, scooter, motorcycle, and bicycle races were held in Italy. Production-racer scooters, such as the 150 Grand Sport, ruled the races, beating all comers—especially the Lambrettas before the TV found speed and reliability—in the early 1960s.

Motorscooter Road Racing

Hanging Off and Rubbing Knees

Put a human being on two wheels, add motive power, and you just have to see how fast you can go. And if you meet another similarly equipped human, you have a race.

Scooterists started racing each other the first day two of them met on a road. The first organized races took place in postwar Italy when a scooter offered the ideal entry into road racing on a budget that could not afford the latest FB Mondial motorcycle or even a resurrected prewar Guzzi racer. Naturally, it was Piaggio and Innocenti who went for each other's jugulars in the scooter class, with occasional challenges from Rumi.

The year 1955 saw the birth of true scooter racing when the French twenty-four-hour Bol d'Or endurance race was staged at the Montlhéry circuit with scooters competing in classes for standard, sport, and racing models. In that first year, a 150cc Lambretta ran in the 175cc motorcycle class finishing 25th overall and first in its class. In 1956, despite legshield-to-legshield competition from Rumi, Lambretta continued to rule.

Off the English coast, the Isle of Man Scooter Rally attracted hundreds of scooter racers in a competition that mimicked the famous motorcycle Tourist Trophy, or TT, race that circled the island. In the United States, NASCAR, the stock car racing association, sanctioned oval dirt-track scooter racing with the premiere held in July 1959 at the New York City Polo Grounds baseball stadium. In typical dirt-tracking stance, the driver would slide the scooter through turns with one foot out.

◄ **1951 Lambretta 250cc Bialbero**
Innocenti's secret weapon, the 250cc double-overhead-cam V-twin racing motorcycle created in 1950-1951. The *V* of the engine ran transverse to the chassis in the style of Moto Guzzi's V-twins. Each cylinder measured square at 54x54mm. Originally beginning life with a single overhead cam on each cylinder, it was modified to dohc to produce 29hp at 9500rpm. Drive to the rear wheel was via a shaft as used on the Lambretta scooters of the time. *Earl Workman Collection*

Lambretta V-Twin Racing Motorcycle

In 1950–1951, Innocenti created a motorcycle—and it was a masterpiece, a 250cc double-overhead-cam V-twin racer. Designed by Ing. Pierluigi Torre, who was responsible for the Lambretta scooter, and Ing. Salmaggi, responsible for the Gilera Saturno and Parilla dohc motorcycles, it was first shown to an awed public at the 1950 Salone di Milano.

Innocenti had aspirations of contesting the 250cc Grand Prix class, but in the end, the 250cc V-twin was run at only a handful of races. Its best showing came in August 1952 at Locarno, where Romolo Ferri was running second to Fergus Anderson's Guzzi Gambalunghino before retiring.

So the Lambretta retreated without ever winning a race. Or did it? Perhaps it had won the most important race it could have competed in, that of showing the rest of the Italian industry that Innocenti could build a motorcycle to be reckoned with should any other major maker think of building a scooter.

In the end, the Lambretta 250cc V-twin was Innocenti's secret weapon.

▼ 1952 Mustang Hot Rod
Salesman-turned-racer Walt Fulton at 100mph on a souped-up Mustang hot rod at Rosamond dry lake in California. The Mustang racers built by Fulton and gang were so fast they were eventually outlawed from Class C racing by the American Motorcyclist Association at the insistence of Harley-Davidson brass after Harley's "real" motorcycles had been beaten once too often by the diminutive scooter. Fulton won races across the U.S. *Michael Gerald archives*

"The sport of racing merits special consideration whether in and of itself or whether for its singular values.... But more efficacious and more exalted is the reality of your symbolic race toward the glory of eternal life. Since you are loyal to the Christian life and you want to conquer not just a trophy that can be passed on to other hands, but a holy, indestructible crown."
—Pope Pius XII to the Vespisti racers of Italy in the 1950s

▲ 1950 Powell Streamliner
In 1950, war was declared between American scooter makers Powell and Mustang. A land-speed-record-Mustang hit 86.12mph; Powell vowed to fight back. In its Experimental Department, Powell's Tony Capanna built a hot-rod P-81 that he believed would top 100mph. He reworked a P-81 engine to 28ci (458cc) run on 10 percent nitro-methane fuel. *Cycle* magazine reported in January 1951 that Capanna set a new class record at El Mirage dry lake: 83.83mph average.

▲ 1949 Lambretta Racer
Innocenti's battle axe was this racer built with at least the idea of the production Model B in the far back of someone's mind. Any parts the street scoot and this full-bore racer shared where purely by accident, but the name on the side of the gas tank was enough.

Shake, Rattle, And Roll

The Fine Art of Scooter Engineering

No one ever blamed a motor-scooter for being technically advanced. Sure, scooters were innovative in their packaging, shoehorning motorcycle components into a dwarf chassis. And yes, they were inexpensive to buy, meaning they were also inexpensive to build due to inexpensive components. In sum, many a scooter could indeed take you from point A to point B—but not all of them could bring you back again.

Consider the engineering history of Cushman. Founded by cousins Everett and Clinton Cushman, the duo had to continually repeat basic engineering courses at the University of Nebraska. Everett finally dropped out; Clinton sweated for seven years to be awarded a bachelor's degree in electrical engineering. Meanwhile, the dynamic duo were building their first Cushman engine in their Lincoln, Nebraska, basement in 1901, an engine that became the foundation of subsequent boat, farm, lawn mower, and scooter motors.

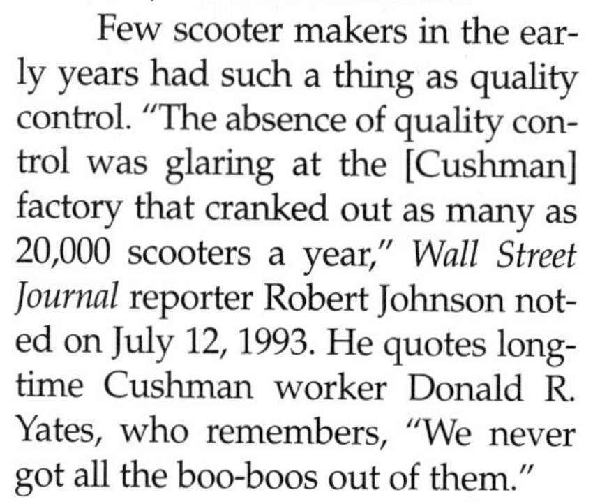

Few scooter makers in the early years had such a thing as quality control. "The absence of quality control was glaring at the [Cushman] factory that cranked out as many as 20,000 scooters a year," *Wall Street Journal* reporter Robert Johnson noted on July 12, 1993. He quotes longtime Cushman worker Donald R. Yates, who remembers, "We never got all the boo-boos out of them."

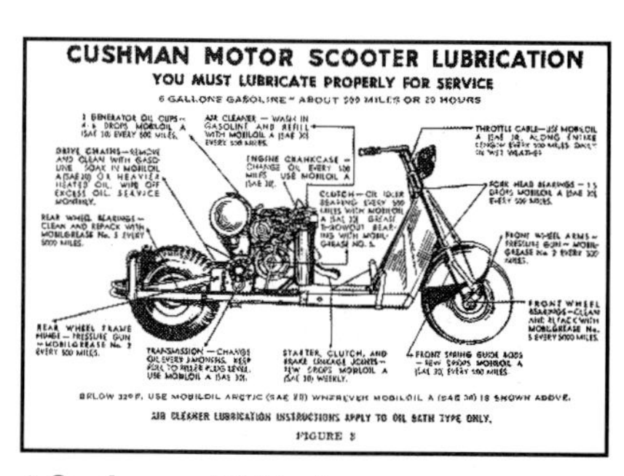

▲ Cushman 50 Series Lubrication Chart
Cushman ads lauded their scooters' properties, ending most claims with at least one exclamation point. But scooter ownership was not for the faint of heart or the mechanically inept: the lubrication sequence alone was daunting and could keep you busy for a full weekend. Launching the Auto-Glide (which was "based" on the Salsbury Motor Glide) in 1936, Cushman ads grandiloquently stated, "Prominent engineers proclaim the Auto-Glide the greatest advance ever made in low-cost motor transportation.... It is under perfect control every second." Sure. Cushman workers still tell of a safety inspector in the Cushman factory who had to jump for his life from a runaway scooter. Enrico Piaggio, meanwhile, was in such a rush to bring his scooter to market that he accordioned years of prototype testing into weeks. Test drivers roared along dirt backroads without air filters on the carburetors; the dust ingested sped up the engine's destruction, and Piaggio engineers did their best to factor in the effects of age. Lo and behold, the Vespa went from prototype to production in just months. *Cushman, Inc.*

▲ Picking Grapes on a 1954 Vespa
Academics and putt-putt pundits look back at the Vespa as a marvel of scooter engineering. It was also a marvel of *economy*. Launched in 1946 by a firm emerging from World War II with its factory reduced to rubble by a B-17's urban renewal project, the Vespa was built of war-surplus parts. Legend has it that the engine was a rehabilitated version of a Piaggio bomber's starter engine, and the novel single-sided front strut was war-surplus aircraft landing gear. The Vespa's most renowned innovation, its steel monocoque chassis-body, was also a brilliant cost-saving invention.

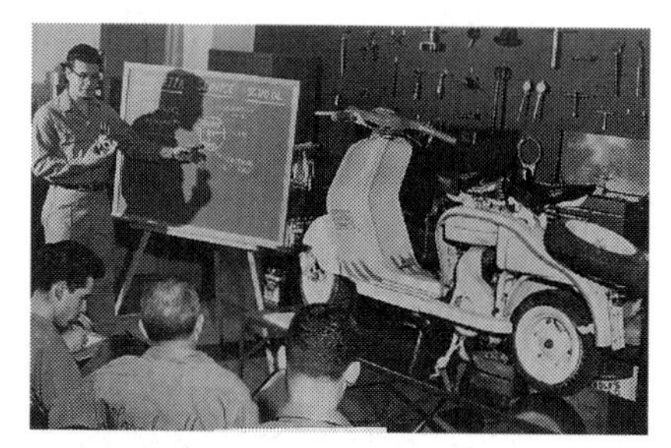

▲ Lambretta Service School
Engines were always the weak link on scooters as they did the most work. The full bodywork hid the motor—and told the story behind the success: Scooters were created as inexpensive transportation for people who didn't want to get dirt under their fingernails or roadspray on their clothes. Thus the 1,000-mile checkup was usually but a dream. Engines worked overtime and got no vacation; they wore out, burned oil, backfired, wouldn't start, seized as solid as cement, or blew up like hand grenades. All of this meant full-time jobs for budding youths in the newly created scooter service profession. Cheap side-valve four-strokes or bare-bones two-strokes, scooter engines were built to be rebuilt. *David Gaylin/Motor Cycle Days*

◄ **Dad and his Salsbury Super-Scooter Model 85**
Pioneer putt-putts eschewed gearboxes, opting for direct drive to the rear wheel, which also saved the tiresome hassles of a clutch. Salsbury introduced its regally named Self-Shifting Transmission in 1937 giving all the ease of an automatic transmission. It was actually a glorified rubber band running on spring-loaded pulleys that contracted with centrifugal motion, thus increasing the pulley-to-pulley size ratio. Nevertheless, any American scooter that was anything had its own "automatic transmission" in no time. Other scooter builders went the more trad route, offering gearboxes with two or more speeds. Usually they worked, other times they jammed, sheared teeth, clutches were sanded down to velvet, or they kicked back like an angry mule. *E. Foster Salsbury Archives*

◄ **1947 Moto-Scoot "Suspension"**
And then there was suspension—or the lack thereof. Few early scooters had suspension, and most putt-putts rode like vibrators on wheels. When Salsbury mounted two hardware-store variety screen-door springs to the front of its Motor Glide, it boasted of a technical marvel as new as tomorrow. And then in the early 1940s, Moto-Scoot bolted on barrel springs, and suddenly the scooter world had dual-rate damping. Even with "suspension," many scooters at speed were notorious for shedding parts like a dog sheds fur. "I'll never forget that special feeling of the wind in your hair when you're going good on a Cushman and then—puff. You just lost a bearing and you look back, hoping to see where it landed," recollects Cushman connoisseur Darrell Ward.

▲ **Lambretta Mechanical Disc Brake**
Yes, Victoria, there were true technical marvels on scooters. Witness the mechanical disc brake that helped the Lambretta TV175 Series 3 of 1962-1965 drop anchor. Yet the high-tech chic didn't come without woes of its own: these brakes are notorious for warping when they got hot, such as by the friction of braking. Owner: Tim Gartman.

▲ **Stranded on the Roadside *mit* Zündapp R204**
All of these technical marvels added up to one big product liability lawsuit—except that there was no such thing way back when. Today, the world's a different place, and product liability lawsuits would muffle the two-stroke cry. Cushman, Inc., still receives letters calling for a scooter revival, according to the *Wall Street Journal*, "They ask why we ever stopped making them," says Cushman spokesman Jerry Ogren. "Today, the product liability lawyers would probably just follow our customers around." *Collezione Vittorio Tessera*

CHAPTER 5

MASS MOBILIZATION

1955-1960

Motorscooters were true democratic machines. Any family could afford one, anyone could drive one, and no class was above them. Even HRH Prince Philip, the Duke of Edinburgh, purchased a fleet of Vespas to putz around the grounds of Buckingham Palace. And to further cement their democratic role, Enrico Piaggio boasted that his machine was the perfect tool to fight Communism.

Scooters mobilized the masses. Horses and donkeys were made obsolete as motorscooters used their puny horsepower to pull farm carts, market wagons, and micro-sized mobile homes. Mere automobiles had never ventured into the cobbled streets of some rural Italian hilltowns perched on mountain sides; scooters broke trail by winding their way through the narrow streets, spreading the news of the twentieth century.

Scooters were a social appliance. With no fear of getting oil on your trousers or burning your silk-stockinged leg on the exhaust due to the side panels, passengers hopped on the pillion seat for a ride. Isolde found Tristan and Romeo eloped with Juliet. Clubs formed to meet others of the same motorized ilk, plan weekend putt-putt getaways, and lament over scooter woes. International rallies were organized where proud scooterists broadcast their fervent nationalism with native customs, folk dances, games, and anything else that could be done on a scooter. Scooters mobilized people like never before.

As a *New Yorker* reporter reported in 1957, "This is more than a fad, it's a revolution and I don't see how anything can stop it." There was more than just a touch of fear in his voice.

"Priests in Italy, according to a Vatican report, currently own 30,850 motorscooters, and in terms of sacraments and good works, the average priest's efficiency has climbed to about 3,000% over that of his road-trudging 19th-century predecessor. Another straw in this high wind is the decline of the more introverted Benedictines and foot-slogging Franciscans in favor of the fast-moving Jesuits, whose high-octane practicality thrives on the motorscooter age. Pope Pius XII has been a longtime friend of automation; last fall he called for 'greater and greater speed to the glory of God.'"
—*Time* magazine, 1952

▸ **On the Road to a Better Future**
Life was definitely good in the late 1950s when prosperity was at hand—or at least in your foreseeable future. The two elfin wheels and hard-working engine of the motorscooter had helped carry the world to prosperity's door.

POPULAR MECHANICS
MOTORIST'S FIX-IT BOOK
Parilla
lo scooter
di chi vuol
primeggiare
ambretta
INNOCENTI
125 li – 150 li
INNOCENTI
motor division
Spare parts list · Catalogue de pièces de rechange
Onderdelenlijst · Reservedelsliste · Reservdelskatalog
Catalogo parti di ricambio · Catálogo de piezas de repuesto
Catálogo de peças sobresselentes · Ersatzteilliste
NSU
PARILLA
Life is Good!

The Vespa–Lambretta War

Absalom! Absalom!

The Vespa-Lambretta War erupted in the time-honored tradition of the Romans and Visigoths, the Montagues and Capulets, and everyone within the Italian government. It started that day in 1947 when Innocenti introduced its Lambretta to battle the dominance of Piaggio's Vespa for the newly created motorscooter market.

The war was fought on every conceivable front—and even a few inconceivable ones. Model specifications, seating setups, paint schemes, road racing and record-setting, owner's club membership, millionth-scooter-built celebrations, and even endorsements from the latest stars and starlets of the day. A Vespisto would not be seen dead on a Lambretta, and for the Lambrettisto there was no doubt that the Milanese soccer teams Milan and Inter could beat Genoa. And if your Vespa sputtered to a halt in Milan's *piazza del duomo*, abandon all hope and run for your life.

▼Model and New Models
The front line in the Vespa-Lambretta War was the showroom, with new scooter models announced almost annually. When the latest Lambretta bowed with a new, neater, or larger gizmo, a brand-spanking-new Vespa was soon out with a bigger and better version of said gizmo. Here, a model introduces the new Spanish Serveta Lambretta models for 1965: the Cento 100, left, and the Slimline 150 Silver Special. You can bet your spare tire that a new and improved Vespa was on its way. *Scootermania!*

◂ Buon Anno, Amici!
And of course there were the annual putt-putt pinups.

▸ Stunt Driver on the Vespa Teeter-Totter
Stunts proved scooter superiority, at least to the faithful followers of each marque. At its annual get-together, Vespa Club of Lisbon members displayed their scootering prowess in time trials and teeter-totter tricks before an awed crowd of fellow riders. The one who won, won. *Scootermania!*

▴ Scooterists and Vespa in Traditional Costumes
Organizing the faithful into owner's clubs around the globe was one way to inspire loyalty to the cause, whether the cause be the Vespa or the Lambretta. Vespa clubs from around Europe then gathered annually for the gigantic Eurovespa hoopla, also sponsored by Piaggio. At Eurovespa 1962, the Vespa Club de España dressed its members and its Spanish-made Vespas in traditional folk costumes to celebrate Spanish culture and Spanish productivity. *Scootermania!*

Scooting With the One You Love

Romance Amid the Two-Stroke Exhaust

Scooters often lead you down the road to love. Whether it's Romeo buffed up by his scooter-inspired virility or Juliet swooning over his Lambretta's chic class, motorscooters are a one-way ticket to that many splendored thing.

Blame it on Gregory Peck and Audrey Hepburn. Ever since the dynamic duo fell in love while touring the Eternal City on a Vespa in the 1953 William Wyler fairy-tale film *Roman Holiday*, scooters and *l'amore* have never been the same. Scooters, once blamed by tourists for raising a ruckus in Rome, were suddenly the vehicle of romance, on the same level with the gondolas of Venice.

Several generations fell in love via motorscooters. While American teens were borrowing the keys to dad's DeSoto, European youths were kickstarting their family's Zündapp to life. Vespas provided the wheels for many a courtship and romance blossomed on a weekend Motoconfort cruise to the country. Cupid rode a Rumi Formichino, and love was in the air surrounding scooterists—if you could breath through the two-stroke exhaust, of course.

► Lambretta Inspires Virility
Frank Sinatra, look out! Nothing wins women like a Lambretta X150 Special, a fedora, and a coolly toted coffin-nail. The bobby-soxers were agog at the well-dressed man with a Lambretta. Presumably the lonely joe in the background has the latest Vespa for wheels—and it probably won't start either.

◄ Want to Go For a Ride in the Country?
Is she wowed by his Heinkel Tourist, or is she laughing? She should be wowed. According to this brochure, the Heinkel is "as streamlined as a jet plane and as smart as new paint!" Facts that should lure many a member of the opposite sex. "No hill too steep, no road too rough, nowhere too far," the brochure also promised; in other words, you need not fear breakdowns in out-of-the-way places—unless they are intentional. *W. Conway Link/Deutsches Motorrad Registry*

▲ A Simson Motorroller and Thou
Lesson number 1 in motorscoot romance: Beware of any suitor with a lesser scooter—especially a moped!—and a cravat.
W. Conway Link/Deutsches Motorrad Registry

◄ Love on a Cushman
The classic Cushman "Turtleback" was the ideal scooter for romance with the add-on "rumble seat" for back-seat gymnastics at the drive-in movie. This 1947 newspaper photo pictured sixteen-year-olds Stuart Crawford and his belle Mary Whitmore on their way to Rocky River High School in Cleveland on their "Chug Bug." As the original caption oozed, "Keeping tempo with the swiftly paced times, school teen-agers are forsaking the foot method for products of the mechanical age. Wizards at flivver finances, the boys and girls choose to buzz about on campus 'teen lizzies.'" Those were the days. . . .

"Sports riders in this country are mostly either single or newly marrieds (scooters are so conducive to romance that there is a fast turnover between these categories)."
—*Popular Science*, 1957

Scooters Across Europe

The Rising Tide of Scooterists

Patriotic and capitalistic firms across Europe rushed to put their citizens on two wheels following World War II. France was the most prolific builder of scooters, alongside Italy, Germany, and England. The first Vespa was constructed under license in France by ACMA in 1950, soon to be followed by scooter creations from Peugeot, Terrot, Motobecane-Motoconfort, and many more.

The Spanish scooter world was typical of many European countries, from Czechoslovakia to Sweden. The firm Lambretta Locomociones SA (later renamed Messrs Serveta Industrial) began building Lambrettas under license in 1952, and Vespa SA was building Vespas in 1962. Montesa created its own bizarre scooter in 1959 that looked like a park bench on wheels and later built the Italian Laverda Mini Scooter under license.

Another route to building scooters was epitomized by the Belgian Cushman firm, created in 1950 to build US Cushman scooters for the Belgians under license. As part of the European Recovery Act, Cushman supplied components including Husky engines, transmissions, wheels, and more to the firm; the Belgian firm made its own frame, forks, bodywork, and other parts.

► *La vie est belle* with a Terrot VMS2 125!
Life is indeed beautiful with a Terrot motorscooter. The beautiful Miss Monde 1953, Denise Perrier, certainly enjoyed "her" Terrot, no doubt given to her by Terrot in exchange for this publicity photo. The Terrot was one of the most unique-looking scooters of all time, resembling a primped and coiffed French poodle on wheels—beautiful, in other words. It was, indubitably, "Une Splendide Réalisation 100% Française." *Bien sûr*, it goes without saying: Who else could build such a motorscooter?

ON THE MOVE WITH "MOBY"

Mobylette

The amazing MOBYLETTE, America's newest, safest method of motor travel. Advanced styling and ease of operation are yours when you own and ride the world famous MOBYLETTE.
Over 250 miles per gallon . . . Perfect for around town and downtown, 30 miles per hour . . . Automatic clutch, all controls conveniently located, makes driving a pleasure for young and old alike.

Moby Scooter

Powerful 150cc engine takes you places quickly, quietly and safely. The MOBY needs no extras to make it one of the world's most deluxe Scooters.
Economical to own and operate. 100 miles per gallon . . . Three forward speeds conveniently located . . . Large 3:50x10" deluxe wheels, extra safe for highways and back roads.

CHAIN BIKE CORPORATION 350 Beach 79th Street Rockaway Beach, N.Y.
Sole U.S. Factory Agents

▲ 1960s Moby Scooter
France's venerable Motobecane firm created a long line of motorcycles and scooters under both the Motobecane and Motoconfort names. For the U.S., the name was Americanized to Moby and you had to read the fine print on this ad to realize that this was a French scooter, which was probably not a big sales point in the land of land yachts. As this 1960s ad huffed, with a 150cc engine, three-speed gearbox, and 3.50x10in wheels, "The MOBY needs no extras to make it one of the world's most deluxe Scooters." Sales in France were huge; sales in the U.S. can probably be counted in the dozens at best.

La scooter Montesa a due posti del 1958

▲ 1958-1960 Montesa Scooter
"Agricultural" would be a compliment to the looks of Spain's Montesa scooter. Perhaps it was a motorscooter designed to a fascist esthetic, debuting as it did during the long "reign" of Fascist Generalissimo Francisco Franco. Whatever, the Montesa scooter was an oddity in a scooter market full of oddities. Laugh if you will, the Montesa was avant garde in construction: Instead of a standard step-through design, Montesa opted to mount two seats atop its fiberglass bodywork, surely some of the earliest use of fiberglass in scooter lore. The hottest features were the seats and luggage rack mounted on rails so they could be adjusted fore and aft. It was an eccentric feature on a scooter that lacked many of the normal features that buyers expected. *Courtesy Bruno Baccari*

▲ 1955 Husqvarna Scooter
"Svensk kvalitet—italiensk skönhet," the Husqvarna scooter was half Swedish, half Italian. Husqvarna purchased scooter chassis from Parilla in Milan, but the Swedish engineers probably did not trust the Italian engines, so they bolted Husqvarna HVA engines in their place and riveted Husqvarna crown nameplates to the legshield. The 120cc two-stroke engine created 4.3hp, and the multicultural scooter rode atop 3.00x12in tires. *Gösta Karlsson Archives*

2 places confortables
2'5 aux 100 kms
Racé et d'un fini impeccable
D'un encombrement minime
Rapide et sûr
Un instrument parfait
La vie est Belle!
AVEC
UN
Scooter
125 cm3
TERROT
Miss MONDE 1953 SUR SON "Scooter TERROT"
PHOTO MIROIR-SPRINT
TRÈS LARGES FACILITÉS DE PAIEMENT
TERROT . DIJON
Magasin d'Exposition:
72, Avenue de la Grande-Armée . PARIS

Scooterists of the World Unite!

Putt-Puttnik Support Groups

▲ **What is it about pork products and scooters?**
These nutty Austrians staged a moveable feast of würst. He who eats the most gets fat. This event at the 1953 annual Austrian scooter meet fell on St. Christopher's Day, so many had their putt-putts blessed by a priest at a religious ceremony. In carrying the extra weight of würst, their scooters needed all the blessing they could get. *Herb Singe Archives*

Sure, torque, brake swept area, and coefficients of drag are always good for a few laughs, but scooter clubs were more than just an excuse for questionable behavior. They became a cultural phenomenon and their meets were social occasions.

Usually these events were sponsored by the parent company, most often Piaggio and Innocenti, to increase sales and encourage acceptability of their wares. During the late 1950s, however, scooter meets became an international event with various nationalities flaunting their native culture and competing for the most impressive getups, always with a scooter, of course. Displays would range from synchronized scooting to dancing the tarantella on Vespas to piling people on a single scooter. These events often lasted for days and eventually reached such a large number that they became a wholesome Sturgis rally.

▲ **1960 Lambretta Raduno**
These sun-soaked legions of Lambrettisti tried to show the world, or at least Italy, that Innocenti ruled the Vespa-Lambretta wars. Their identical steeds are the recently released Series 2 Lambretta TV175s, to which Piaggio would respond in 1962 with their speedy Vespa 160 GS. *Collezione Vittorio Tessera*

▲ C'mon join the fun!
It's Hawaii Day in balmy Kent, England, as The Innocents parade around town, which is sure to convince nonmembers to join up. In the 1950s, the local Lambretta dealers would sponsor events like this and hope for as many participants, knowing that the publicity made the cash register ring louder. *Mick Walker*

> "It's an unwritten code for scooterists to greet each other."
> —Scooterist quoted in *New York Times Magazine*, 1958

◄ Buzz, buzz, buzz
How can you have a club without an emblem to let others know that they aren't in the gang? These Vespa club logos are a few examples of the Piaggio-inspired designs.

Boltstrippers' Anonymous
"Ne'er too tight is a bolt."

As part of the revival of scooter clubs in the 1980s came scooter 'zines, small newsletters dedicated to the cause of putt-putt resuscitation. Everything from scooter want ads to the upcoming skafest to Mod philosophy were debated between their covers. Any submission was welcomed.

One such 'zine, *Boltstrippers' Anonymous*, came from the depths of the Breath of Exhaust Lambretta Detuning Works and dedicated itself to the prevention of rounded nutheads and stripped bolts everywhere, with such highfalutin' messages as, "I say no! No to these scooters rotting away in the dark abyss of a dust-drenched garage bespeckled with devilish rust. I regard my image in yon rearview mirror and see not merely a restorer of Italian scooters, but one who is in the business of saving souls!"

Factories as Icons

Epiphany in Steel

In the 1930s and 1940s, recovery from the Great Depression was epitomized by factories spewing out smoke from their chimneys. In the 1950s and 1960s, scooter ads heralded the factory as an icon of progress, cleanliness, and efficient mass-market technology. The production line reigned supreme as photos showed line-ups of row upon row of scooters to prove that quantity equals quality.

The number of scooter factories that survived in Europe past the 1960s can be counted on two hands. While some factories were able to sell off their tools and dies to Third World countries, others successfully shifted with the market and produced miniature cars or anything else that was selling. The Piaggio factory outside Genoa was able to hold on and keep producing Vespas, while their arch-rival Innocenti moved to producing Innocenti (Austin) Minis. Meanwhile the other European motorcycle manufacturers that had dabbled in scooters stuck with what they knew best and covered up the scooter skeletons in the closet, claiming it to have been but a passing fad.

▲ Lambretta Assembly Lines
Enough to make any Lambrettista green with envy: every 60 seconds a brand new Lambretta Slimline rolled off the assembly line to a happy new owner. "There's a color or color combination just for you! Match your personality from the wide range of available colors," exclaimed the ads. As if the 6.6hp of the Li125 wasn't enough. *David Gaylin/Motor Cycle Days*

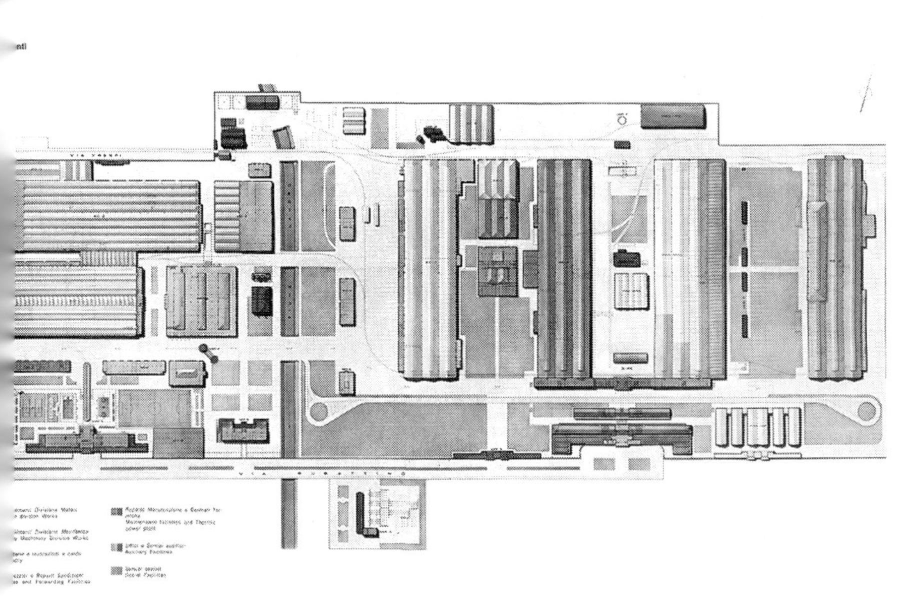

▲ **Workers of the World, Unite!**
After a hard day's work in a faceless Bauhaus-style factory, every worker can go home with a smile as long as they're on a Bastert. These speedy scoots topped 50mph for the 174cc version and 55mph for the 197cc scooter. Unfortunately, only 1,200 of these *elegant and schnell* two-wheelers, whose ads boasted that the design was "outstanding perfection," were built. Although the Einspurauto (meaning "Single-Track Car") may have been curvaceous with an elegant dashboard and styling, the factory from whence they came lacked the sumptuous lines that made this scooter a gem.

▲ **Birthplace of the Lambretta**
Located on the bubbling Lambro river in the beautiful Lambrate section of Milan lies the glass-and-cement buildings that pumped out the scooter whose name rang on everyone's tongue, Lambretta. While Leonardo Da Vinci had designed the extensive canal system of Milan, he didn't allow space for one essential feature that the designers of the Lambrate factory knew had to be included in any Italian design, a soccer field. This factory was immortalized in the film *We Carry On*, which idealized the massive production line while the industrial noise/music of Musique Concrète played in the background. *Collezione Vittorio Tessera*

▼ **UFO Showroom**
This tidy array of scooters in the Salsbury factory is enough to make even the most stoic scooter aficionado drool in lust. This line-up of prototypes would rocket these space-age Model 85 Standards and Deluxes into the scooter hall of fame. *E. Foster Salsbury Archives*

Scooter Raids!

Around the World on a Putt-Putt

Once on a scooter, one gets bold—often too bold. No mountain is too high, no desert too hot, no distance too great, and thus raids began. Thinking their shiny new scooter capable of anything, crazy youngsters would travel ridiculous distances driving straight through the night just to show that nothing could stop them.

The final destination of raids was often a scooter rally or club meet, gatherings of the only people who would truly appreciate that someone just drove hundreds of miles on a vibrating two-stroke. Tales of raids range from circling the globe to conquering the jungle to ping-ponging across the U.S. any number of times.

▲ Running Away Forever–or at Least a Day
Armed with only a Cushman, these two fourteen year-olds planned their trip through Texas and Louisiana with not a fear in the world. In 1947, all one needed was a map and a knapsack and confidence that the scooter's engine would not seize in the hot southern sun. *Herb Singe Archives*

HEINKEL
Tourist

No hill too steep,
No road too rough,
Nowhere too far...

◀ Only Kryptonite Can Stop You Now
"HEINKEL TOURISTS have covered hundreds of thousands of miles in Australia, Africa, South America and in other tropical regions under the most adverse conditions," reads the ad copy. So don't have second thoughts about road-tripping your Tourist in the mild climates of the U.S. and Europe. And what's more, "the machine runs on ordinary pump petrol—all you have to say to the garage-man is 'fill her up.'" *W. Conway Link/Deutsches Motorrad Registry*

◂ The World's Your Backyard
If the Kreidler-Roller can get from as far as Munich to the Suez Canal through the Orient to the American continent, what are you waiting for? Advertising like this pushed the possibilities making just a trip to the grocery store practically an insult to the scooter. One of the advantages of traversing the globe on such a small 50cc engine is the relatively simple repairs. *W. Conway Link/Deutsches Motorrad Registry*

"No hill too steep,
No road too rough,
Nowhere too far . . ."
—Heinkel Tourist advertisement

◂ Ohio—Or Bust!
No, he's not trying to set the record for the most heavily laden motorscooter. Howard J. Bradley is merely heading 140 miles to a scooter swap meet in Bucyrus, Ohio, with 160lb of equipment on his trusty P200E Vespa.

Scooters and Women's Emancipation

Riding Side-Saddle No More

"I bought a brand new Lambretta 150 for $300 and put on thousands of miles driving around Manhattan on weekdays and out to Fire Island on weekends. En route, I discovered that the scooter provided a reliable litmus test of male character. Those who were threatened by it didn't last long in my affections."
—Letty Cottin Pogrebin, *Ms.* magazine, 1987

◂ **Riding Side-Saddle on a Vespa 160 GS** Side-saddle was of course the way a woman should ride a motorscooter, the true ladylike style. But all of that was about to change, due in a large part to the freedom and wider vistas the scooter opened up to women in Europe. Suddenly, women had mobility, which brought freedom, which brought wider horizons, which eventually brought change to women's status.

"The narrowing of the new-look skirt was dictated in order to prevent it getting tangled up with the wheels. The slipper shoe was created for footplate comfort. The turtleneck sweater and the neckerchief were designed against draughts down the neck."
—*Picture Post*, 1950

▾ **Side-Saddle Accessory**
A 1957 catalog of Vespa accessories proudly displays the *Poltrovespa*, or Vespa easy chair, a side-saddle for women with all the glamour of a corset.

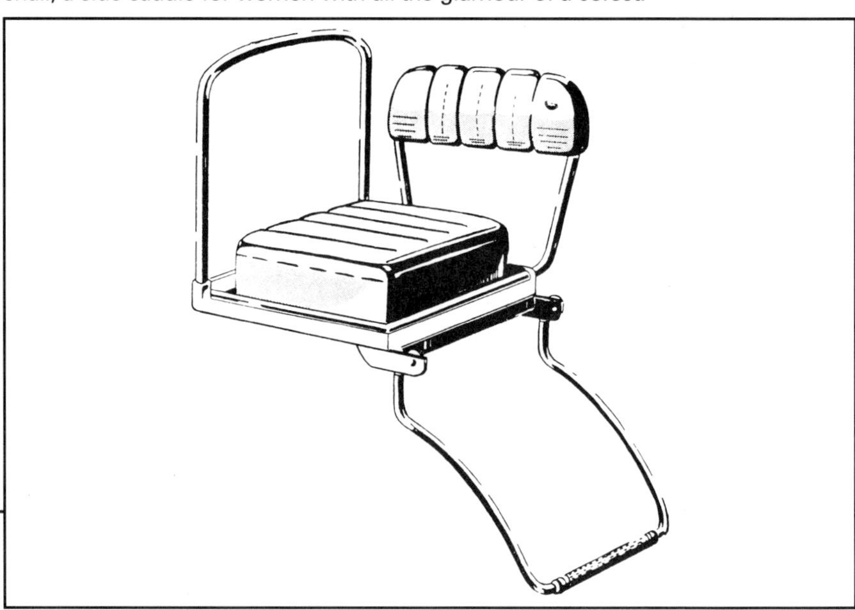

Motorscooters played a starring role in the emancipation of the modern Italian woman. The British magazine, *Picture Post*, documented "A New Race of Girls" in 1954, stating that "the motorscooter gave her new horizons," as quoted from Dick Hebdige's semiotical analysis of the scooter in *Hiding in the Light*. A 1954 Innocenti promotional film for England, entitled *Travel Far, Travel Wide*, showed an airplane stewardess (in itself an image of the modern woman) on her Lambretta. The narrator announced that "The air hostess can become the pilot herself—and there's plenty of room on that pillion for a friend!" And who should climb onto the pillion but the airplane's male pilot.

Along with freedom of mobility, scooters were a catalyst in changing female fashion, symbolic of changes in status. *Picture Post* even went so far as to say that the scooter—along with beauty competitions and films—"consummated" the Italian woman's emancipation. As proof it pictured film stars Anna Magnani and Sophia Loren and talked about a new breed of "untamed, unmanicured, proud, passionate, bitter Italian beauties"—all riding scooters. Italian women had come a long way from sitting side-saddle on the pillion.

▲ Mod Woman and 1947 Lambretta Model A
Buttoning up her leather jacket before blasting off on her brand-spanking-new Model A Lambretta, this woman had come a long way on two wheels. As many 1950s style magazines proclaimed, even hairdos changed in response to women's access to scooters: Headscarfs to protect your long locks gave way to short hair as the ideal style for wind-in-the-hair riding. But all of this did not come as easily as simply kicking over a scooter engine: Innocenti's prudish *Lambretta Notiziario* owners' magazine lamented that "one is all-too-frequently tormented by the sight of badly trousered women on motor scooters." *Collezione Vittorio Tessera*

▲ Classy Lady and 1955 Triumph TWN Countessa
Soon, women were piloting their own scooters. Scooters were designed with women in mind, from the first ABC Skootamota of 1919 to the premier Vespa. The enclosed bodywork protected your clothes from roadspray and inclement weather, and the open step-through design facilitated dresses and ever-shorter skirts. The engine was reliable and the mechanicals were covered, meaning they wouldn't need adjustment and get dirt under your fingernails. Controls were simple to use if you had never driven a motorized vehicle before. And floorboards were *de rigeur* instead of motorcycle pegs, which were hell in high heels.

The Luxurious Motorscooter

Cruising the Piazza in Style

In 1952, Moto Ducati launched a new concept in scooters with its Cruiser. Here was utilitarian transportation but with stylish luxury, the first *gran turismo* scooter, in the style of Enzo Ferrari's new GT automobiles. Ducati designers reasoned that the economy was improving and people would want to move upscale, but still could not afford a car. But the Cruiser was too much too soon, and it was gone by 1954.

But the *gran turismo* idea behind the Cruiser did not fade away. Inspired by the luxury scooter concept, other scooter makers would launch their own luxury models within the coming decade.

▲ 1952-1954 Ducati Cruiser
The luxury motorscooter was born when Ducati launched its fabulous Cruiser scooter at the Milan Fiera Campionaria of 1952, boasting a list of features far beyond any scooter of the day. The Cruiser looked like a zoot suit on wheels; it was draped in elegant, flowing two-tone bodywork penned and produced by the celebrated Italian carrozzeria, Ghia. The long right-side cover swung open effortlessly on a front-mounted hinge to reveal the 7.5hp 175cc overhead-valve engine with the cylinder set transversely in front of the rear wheel, 12-volt electrics, electric start, and an automatic transmission based on a hydraulic torque converter. With the powerful four-stroke engine, the Cruiser actually had to be *de*tuned to meet the Italian government's maximum scooter speed limit of 80km/h. Suspension up front came from telescopic forks with a hydraulic damper mounted at the hub; the rear wheel was held by the massive aluminum-alloy gearbox case that doubled as a swing arm with a hydraulic damper mounted horizontally below the engine. Ducati was partly funded by Opus Dei, the Vatican's investment arm, but even with Papal blessing, the luxury Cruiser only remained in production two years. *Collezione Vittorio Tessera.*

◄ 1957 Cosmo Scooter
Cosmopolitan Motors of Hatsboro, Pennsylvania, sold its own badge-engineered Cosmo Scooter in the mid-1950s. Built by Puch of Austria, the Cosmo was the Cadillac of motorscooters, bedecked with a vicious red paint, quaint two-tone saddle, windscreen, spare wheel, front and rear luggage racks, and acres of chrome doodads. The only thing it lacked was tailfins. Never mind. This was high scooter style circa 1957.

▲ 1950s NSU Prima Dashboard
The array of gauges, buttons, and knobs made you feel as if you were piloting a jet aircraft—all on a relative scale of course, when you're talking motorscooters. Still, while the NSU Prima may not have reached mach speeds and broken the sound barrier, the dashboard was the coolest of the cool. Here you had a clock, speedometer, odometer, choke, key to the electric ignition, battery warning light, grocery bag holder, and glovebox. The Cadillac of the day didn't even have a grocery bag holder!

◄ 1950s Maico Mobils and Sun Goddesses
Motorscooters were by definition enigmas, but the Maico Mobil was an enigma among scooters. These two sun goddesses *cum* housewives were on their way to market with their two-wheeled grocery getters when they stopped to chat, exchange gossip, cookie recipes, and gas mileage comparisons. Their Maico Mobils were among the coolest scooters ever built. These two women look like they knew it.

The Vespa Gran Sport

The Scooter that Transcendeth All Knowing

Piaggio and Innocenti also took heed of Ducati's *gran turismo* Cruiser scooter. By the mid-1950s the time was right for a luxury scooter. Italians had extra money in their pockets: That extra money could buy a more stylish scooter with extra power and luxurious accessories that could be driven to work during the week and be loaded up for a trip to the sea or a picnic in the country on weekends.

In 1954, Piaggio beat Innocenti to the punch, launching its Vespa 150 Gran Sport, a refined and perfected high-performance touring and sporting model. Innocenti responded in April 1957 with its Lambretta Turismo Veloce 175. Again the two scooter giants were locked horn to horn in competition.

> *"Paradiso per due."*
> —Vespa Gran Sport flyer, 1960s

▲ 1962-1964 Vespa 160 GS
Paradiso per due, or "paradise for two" as this early 1960s Vespa GS brochure promised. In 1957, Innocenti had responded to the Vespa Gran Sport by offering its Lambretta TV 175. With a 175cc competitor threatening to steal its two-stroke thunder, Piaggio created its 160 GS of 1962-1964 by enlarging both the bore and stroke of the 150 GS engine to 58x60mm. As this brochure stated, "Speed, acceleration, stability, comfort, elegance, economy, these are the qualities of the Vespa GS that emerge in every condition of employment." Couldn't have said it better ourselves. *Collezione Vittorio Tessera*

▲ 1955-1961 Vespa 150 GS Racer
When Piaggio debuted the Gran Sport at the Salone di Milano in late 1954, it must have known they had created something special. The GS stunned everyone. It featured a new engine with a square 57x57mm and 6.7:1 compression ratio, pumping out 8hp at 7500rpm via a UB 23 S 3 Dell'Orto carburetor. Backed by a four-speed gearbox, the GS reached 62mph on new 3.50x10in wheels. The GS pictured here in Milan alongside a Team Vespa racing team VW Transporter was modified for racing with aerodynamic wheel covers and the early-model spare tire stored on top of the central tunnel, as the left sidecover now housed a luggage trunk. The 150 GS came with a bench seat for two people and was available exclusively in metallic grey. With the GS, Piaggio had refined and perfected its scooter into a high-performance touring and sporting model that is still today considered as perhaps the best scooter ever.

▲ 1964 Vespa 160 GS
The cool scooter and typical tough, leather-jacketed scooterist of the mid-1960s. By the middle of the decade, Piaggio sensed that the market for scooters had shifted: Sales of utilitarian scooters were being displaced by mini- and full-sized cars, and scooters were becoming a cult symbol of the growing youth market, winning an image as a mode of transportation primarily for teenagers before they could afford a motorcycle or four-wheeler. Piaggio responded in 1964 by enlarging the 160 GS engine to create the 180 Super Sport, which replaced the GS as the top-of-the-line Vespa until 1966. The 181cc engine was based on 62x60mm, creating 10.3hp at 6250rpm. The 180 Super Sport obtained a new body-chassis, straying from the classical rounded styling of the GS to the angular look of the mid-1960s. The SS was superseded by the 180 Rally of 1968, which was in turn followed by the Rally 200 and Rally 200 Electronic of 1972–1977.

Lambretta Turismo Veloce

Horsepower Corrupts Absolutely

▲ 1957–1958 Lambretta TV 175 Series 1 and LD125
TV stood for Turismo Veloce, or Touring Speed, a scaled-down translation of *gran turismo* into scooter-speak. Innocenti's TV1 was a completely new design from the wheels up when it debuted in 1957. The 170cc two-stroke engine measured a square 60x60mm and delivered a powerful 9hp, significantly more than any earlier Lambretta or Vespa. A fourth gear was shoehorned into the transmission, and the shaft-drive of all earlier Lambrettas was shelved in place of a modern enclosed duplex chain drive that required no adjustment, lubrication, or cleaning. Wheels were now 10in with 3.50x10in Pirelli tires. The TV1's bodywork enveloped the scooter like fluid metal; the curves followed the scooter's function with rich expanses of flowing steel that made the new Lambretta look more modern than Piaggio's GS, which only a trained eye could differentiate from the first Vespa of 1946. But there were problems looming. The TV concept was right, the timing was right, but Innocenti's execution was wrong. Whereas Piaggio's GS became its best scooter of the 1950s and 1960s, the first series TV 175 was a dismal failure. It was underdeveloped and won a reputation for poor reliability that haunted Innocenti into the 1960s. Never mind: In January 1959, the redesigned TV2 would be a worldbeater.

The Piaggio-Innocenti Wars flared in the mid-1950s on the luxury scooter front when the Vespa GS was released in 1955. Suddenly Vespa ruled the roads. But it couldn't last, and it didn't. In 1957, the Lambretta Turismo Veloce was launched as a 175cc high-powered rocketship—25cc and 1hp more than the GS could muster.

The first TV, however, was a failure. Rushed into production, it won Innocenti a reputation for unreliability that the firm fought to live down throughout the 1960s and 1970s (it was probably the Vespisti who spread the rumor).

But never fear. In 1959, the TV Series 2 bowed with a new engine based not on the old TV1 but on the highly successful Li125. With the TV2 and the subsequent TV3 of 1962, Innocenti was crowned king of scooterdom. If power corrupts, then horsepower corrupts absolutely.

◄ 1962-1965 Lambretta TV 175 Series 3 Slimline
In 1961–1962, the bodywork of the Li and TV was redrawn. The new Slimline styling was sleek and angular with flash replacing the fleshy look of the earlier Lambrettas. The Slimline looked like a Lambretta on a diet; the curvaceous Marilyn Monroe styling that characterized the 1950s Lambrettas was shed in favor of the thinner Twiggy look of the 1960s. The TV3 wore the new go-fast Slimline styling over a high-performance 8.7hp engine and front hydraulic dampers for a smoother ride. With the TV3, a 20mm Dell'Orto replaced the late TV2's 21mm. The TV3 dropped anchor with a mechanical disc front brake that pressed a full-circle pad against the rotor. It was not the first use of a disc brake on a motorcycle but it was the most influential, leading the way for motorcycle and automakers in the mid-1960s. Under the old hot-rodder's dictum that there's no replacement for displacement, the 175cc engine was increased to 200cc for the TV 200 model, aka GT 200, which lasted until 1965.

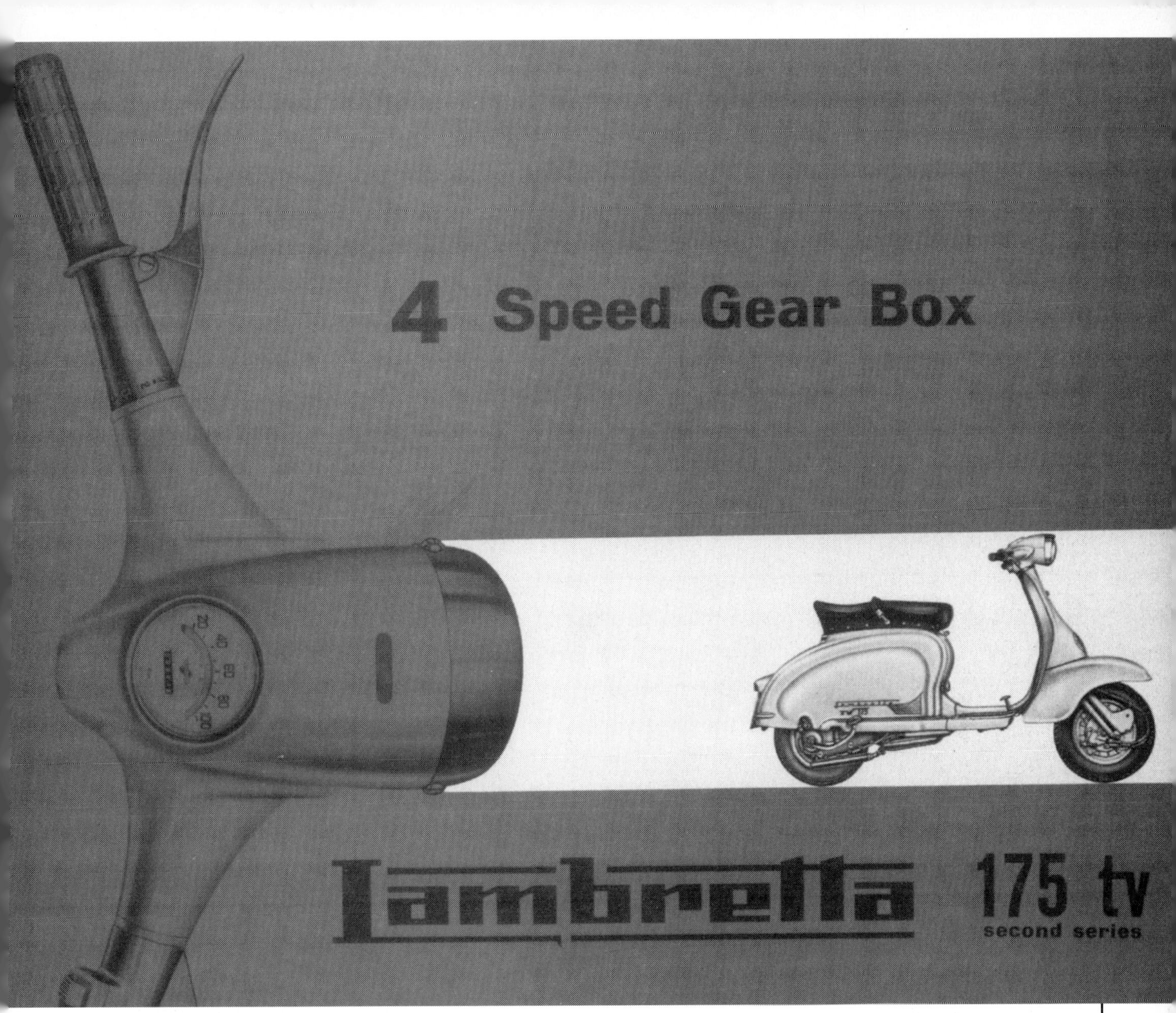

▲ 1960 Lambretta TV Series 2
Funky cool brochure graphics as only the Italians could dream of promoted the key identifying feature of the Series 2: The headlamp moved to the handlebars to cuddle beneath a streamlined cowling that also incorporated the speedometer unit. But there were many more changes under that exquisite bodywork. After just sixteen months of production, the TV1 was junked to be replaced by the TV2, which was a new scooter based not on the TV1 but on the phenomenal new Lambretta Li125 of 1958. The only items carried over from the TV1 were the name and the body style, but that too was updated. The new 175cc engine was a larger-bore version of the Li125 engine at 62x58mm, creating 8.6hp. With the TV2, Innocenti had its flagship sporting scooter that it should have had in the TV1. The TV2 would receive a facelift in 1962 with the introduction of the new Lambretta Slimline styling, and carry on to 1965. *Collezione Vittorio Tessera*

▲ Lambretta Display Truck
Packed with new TV1 scooters, this display truck toured Italy to show off Innocenti's wares in a style perfected by Italian fish vendors. The sides of the truck expanded outward to showcase this mobile showroom. *Collezione Vittorio Tessera*

Scooter Stunts and Putt-Putt Promotions

Manifest Destiny, Scooter Style

▲ Courtesy is Contagious
In the 1920s advertisers paid people to wear advertising boards; in the 1950s, they paid people to drive scooters. This American beauty would cruise the city in her teardrop-paneled TWN Tessy passing out good driving awards and a smile. So much for drag-racing your scooter. *W. Conway Link/Deutsches Motorrad Registry*

As an alternative to the automobile, scooters hit the market with such force that nearly every motorcycle company considered going into the lucrative scooter biz. Producing one, however, was never enough; publicity and outrageous promotion was the key to assure success. Shameless gimmicks promoted everything from scooter sports, like polo and bullfighting, to drivers' education schools and scooter service stations.

Apart from posters and advertisements, publicists presented photo ops showing a new class of young people interested in cheap transportation, meeting fellow scooterists and being on the cutting edge of style. You too could join this hip avant garde club—if you could afford a scooter. You could use their special mechanic shops, participate in their rallies, and try any range of dangerous or goofy stunts on your scooter that would be written about in their next newsletter. If you owned a scooter, you belonged.

► A Driver's Ed. Horror Film in the Making
Every driving instructor's nightmare: driver's ed. behind the handlebars of a scooter. At least with this scooter, the handlebars could be controlled by the back-seat driver, making this the first tandem scooter. This Lambretta Li125 was specially designed for driving schools like this one in London. If learning to drive a scooter was rough, perhaps taking the bus would make this a safer world.

Our mightiest weapon
–your mailed letter!

Write ere the sun sets! Declare your allegiance to the League of Honest Coffee Lovers!

Since our crusade for honest coffee, our ranks have grown a hundred-fold. Those who work to deny your right to an honest cup of coffee—*coffee made with one Standard Coffee Measure of coffee to the cup*—turn even more pale as our phalanxes grow.

Now, a nation has been alerted. Our men stand firm. Our women pay no heed to the tempting voices that would have had them spare the coffee and spoil the cup. In home after home the pungent aroma of honest coffee proclaims our success.

Are you numbered among us? Declare yourself—lest your moment of glory slip by. Take your pen now and write "Aye!" that we might mark you among us.

JOIN THE CRUSADE FOR HONEST COFFEE

Write in today for your complete Honest Coffee Lovers' Kit. It contains everything you need: "The Secret of Honest Coffee," the official Standard Coffee Measure and a framable Certificate of Membership in the League of Honest Coffee Lovers. Send your request to Pan-American Coffee Bureau, P. O. Box 28, Old Chelsea Station, New York 11, New York, and please enclose your initiation fee of ten cents.

LEAGUE OF HONEST COFFEE LOVERS

◂ Scooter Crusading for the League of Honest Coffee Lovers
Scooteristi have always hung at the coffeeshops—even back in 1959 when the League of Honest Coffee Lovers (LHCL) crusaded for more java in each cup of joe. As this ad from *Life* magazine proclaimed, "Now, a nation has been alerted. Our men stand firm. Our women pay no heed to the tempting voices that would have them spare the coffee and spoil the cup. In home after home the pungent aroma of honest coffee proclaims our success." Explain all this if you will. The 1950s were odd times.

▴ Scoot-a-Muck
Although the design of enclosed scooters resembles medieval covered steeds, these jokesters in a Berlin suburb in 1955 carried that idea a step further by replacing the horses altogether leaving scooter chariots. Unfortunately, four feet allow for much better traction in mud than wheels, whereas these drivers just slid around the track, no matter if the scooter was a Vespa or Lambretta. *Herb Singe archives*

◂ Are Even Scooter Customers Always Right?
Once upon a time, scooterists expected nothing less than top-notch service. They ruled the streets with two-stroke exhaust, and service stations knew that if they wanted to sell a half-gallon of gas, they better check those tires.

Weekend Warriors

Putt-Putts Escape Package Tours

The idea of getting away for the weekend was finally made a reality to anyone who could afford a scooter. Here was a way to get out of the crowded cities to the crowded beaches. But now, vacationers were no longer boxed in by specific destinations of trains, but rather they could find the exclusive cabin to get away from it all. Nevertheless, beaches and mountain getaways were swarmed with tourists on buzzing Vespas in search of the perfect tan. Or at least if you could get to a vacationing hot spot, the ambitious tourist could simply rent a scooter for the weekend and zip away from the crowds.

◄▲On the Way to Sun and Fun
Innocenti pushed their powerful scooters as the perfect transportation to get away from it all. No more jamming into crowded trains or buses to get to the beach when you've got "Lambrettability." While Piaggio's ads often focused on the practical aspect of the Vespa, Innocenti pushed the Lambretta as more of a luxury mobile letting you live the good life. Since you wouldn't be caught dead in an Italian city during August, Innocenti threw in five free liters of Agip gas when you picked up your scooter.

▲ Family Outing, Vespa-Style
This Portuguese nuclear family has the right idea of hitting the beach on a hot, sunny day in 1954. Although the price of these two 125cc Vespas would be comparable to a mini car, who wants to travel in a cramped auto when you could feel the wind in your hair (blowing your hat) and getting a tan to boot? Unless of course it's raining. *Scootermania!*

◄ Hunting Lambretta
Maybe this Series 2 Lambretta would make a good decoy for attracting deer because the banging two-stroke engine would scare away any living being within miles. It may be perfect for tracking already wounded game through the woods, but a little cumbersome for attaching the kill to the front fender. *Collezione Vittorio Tessera*

The Evil Scooter Menace

Beware the Unwary Tourist....

By 1956, 600,000 scooters were buzzing along Italian roads. Tourists to golden Italy, however, had little respect for the scooter's role in mobilizing the reconstructed country; instead, the camera-toting hordes despised the unpicturesque sight of the Vespas swarming like insects around the Trevi Fountain, blurring their snapshots for the folks back home and filling the air with the choking scent of two-stroke exhaust.

This stupidity continues to the present day in such trendy snobola "style" books as *Italian Country* by Catherine Sabino (Potter, 1988), which describes a quaint tourist's view of Umbria: "Passageways an arm's length in width will never hear the rumble of an automobile, or with any luck, the irritating buzz of the Vespa."

▲ Waiting for You in *Roma*
Love—and a 1954 Vespa 125—await you in Rome along the banks of the Tevere, Castello San Angelo in the golden background. Even the hugely popular film *Roman Holiday* could not save the scooter's bad rap. The 1953 William Wyler flick featured Gregory Peck as the suave American journalist and Audrey Hepburn as the cloistered princess who find love aboard a Vespa in Rome. The movie lifted the American public's image of a scooter from an insect to at least the quaint level, but damned if the typical American tourists were actually going to ride one of those glorified Eye-talian lawn mowers around the Eternal City.

"Among Italy's contributions to civilization, the motorscooter cannot be counted as an unmixed blessing. A visitor to Florence, for example, may be lulled to sleep by the strains of *La Tosca* emanating from the municipal opera house, but he is in for a rude awakening when the opera lets out. Indeed, the noise made by a thousand homeward-bound Florentines, most of them riding motorscooters that sound like riveting guns, is enough to drive a tourist back to New York for a little peace and quiet. Now it appears that Americans won't be able to escape motorscooters even by staying home. Not content with making the Italian night hideous, Piaggio & Co. has launched a determined assault on the American market. The prospect is in a literal sense disquieting."
—*Fortune*, August 1956

▲ Roman Holiday—Ruined!
You had just aimed your Kodak Brownie for another out-of-focus shot of St. Peter's when one of those pesky little Vespas zipped in front and ruined your photo for the folks back home! No wonder they're named after insects! *Herb Singe Collection*

▸ *Polizei mit* Vespas
It's a conspiracy! You went to the Old Country to see the quaint sights, and those darn scooters are everywhere—why even the police ride them! These two Düsseldorfer troopers should be justifiably proud of their Hoffman Vespas with the spiffy searchlight mounted on the handlebars. Cop lights, cop tires, cop two-stroke 125cc engine. . . .

▲ Heinkel Tourist on Tour
Not content to rattle all the windows on the *strasse* in their hometown, this Dutch couple even brought their three-wheeled scooter menace to America! All part of a fourteen-month around-the-world honeymoon trip via scooter, according to Jan and Leny Hoe. After crossing America, the duo were on their way to Australia, Ceylon, India, Pakistan, Iran, Turkey, and home via European byways. If their marriage could survive that, it could survive anything. *Herb Singe Collection*

Motorscooters by US Mail

"Some Assembly Required"

▲ 1946 Beam Doodle Bug and Gambles Stores Sign
In 1946, Gambles signed the Beam Manufacturing Company of Webster City, Iowa, to build the Doodle Bug. Beam chose the tried-and-true 1 1/2hp Briggs & Stratton engine to power the Doodle Bug via a belt drive, although for a short period, the B&S engine was out of stock and Beam filled in with a Clinton 1 1/2hp, making the Clinton-powered Doodle Bugs rare insects. The Webster City factory hatched Doodle Bugs in four build lots of 10,000 scooters each for a total of 40,000. Owner: Jim Kilau.

Down on the farm or in small-town America far from the centers of commerce, the concept of mail-order shopping blossomed after the turn of the century. Led by catalog retailers such as Montgomery Ward and Sears, Roebuck & Company you could mail or telephone in your order for everything from a new frock for mom or a suit for pa to a full-size upright piano for junior—farm tractors and pre-fabricated houses and barns were even offered!

> "All-new fresh-as-a-breeze styling! All-new distinctive colors! Shipped partially assembled in wooden crate."
> —1959 Sears, Roebuck & Company catalog ad for Allstate Jetsweep

Motorscooters were available by mail before World War II through *Popular Mechanics* ads, but when the Gambles Stores began selling Doodle Bug scooters in 1946, mail-order motorscooters took on a whole new meaning. Suddenly, thousands of putt-putts were being sold through catalogs by Gambles, "Monkey" Ward, and Sears, Roebuck & Company. Large profits were reaped from the little scooters, and the Doodle Bug, Cushman-Allstate, and Ward's Riverside motorscooters became household names.

► 1946-1948 Doodle Bug Ad
Roy Rogers rode Trigger, Dale Evans saddled up Buttermilk, Gene Autry hotspurred Champion, the Lone Ranger and Tonto had Silver and Scout respectively, but the elfin Doodle Bug scooter became the dream of many a city-bound young American buckaroo in the 1940s. "Doodle Bug the real deal," promised this rare Beam ad for the "Xtra new" model. *Herb Singe Collection*

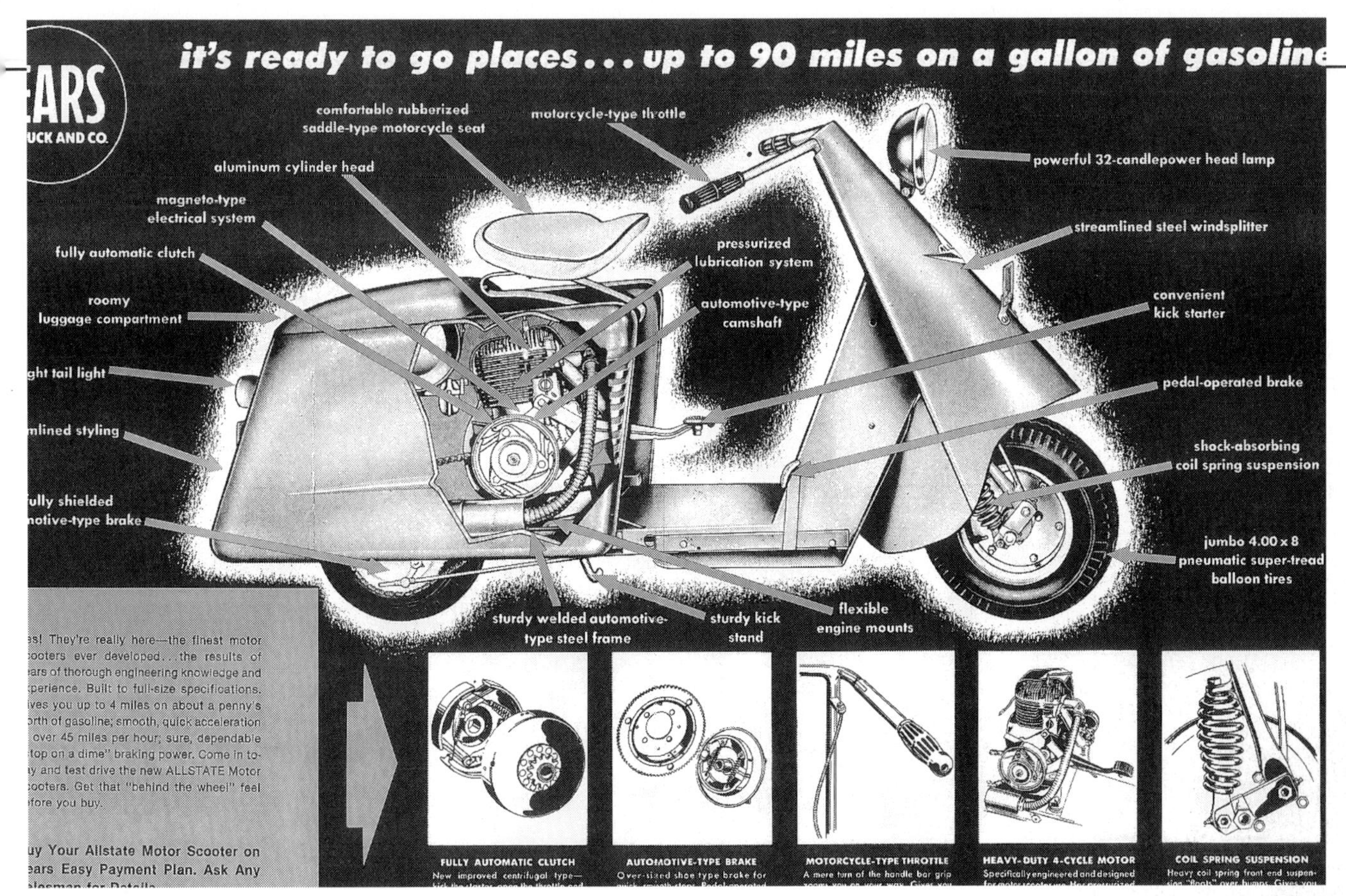

▲ 1951-1957 Sears Allstate Deluxe 4hp Model
"Yes! They're really here— the finest motorscooters ever developed . . . the results of years of thorough engineering knowledge and experience," proclaimed this Sears ad. The engine was the famed 14.9ci Husky; Sears ads ballyhooed its "automotive-type camshaft," which was certainly superior to other camshaft types. The rear drum brake was also of the "automotive-type," giving "Dependable 'stop on a dime' braking power," in technical terms. "Runs easily in slow traffic, zooms up to 40 or more for speedy get-away," the catalog stated. The "or more" top speed was 45mph with a tailwind.

◄ 1961 Riverside Lambretta Li125
Montgomery—better known as "Monkey"—Ward carried Lambretta scooters, badged-engineered as Riversides. This 1961 Slimline Li125 was made into a Riverside with an identity crisis by merely riveting on a new chrome emblem to the front legshield and slapping on an instruction sticker in English. Ward also sold Mitsubishi Silver Pigeons, which had been based on the Lambretta LD125. The Silver Pigeons were imported by the Rockford Scooter of Rockford, Illinois, and labeled as "Rockfords" with no mention of "Made in Japan" so soon after the war. Ward renamed the Lambretta-Mitsubishi-Rockford line as the Riverside Nassau, Waikiki, and Miami. Owner: Eric Dregni.

▲ 1951-1957 Sears Allstate Deluxe 4hp Model
Sears, Roebuck and Co. saw the success of Gambles' Doodle Bug and yearned for a scooter to call its own. In 1951, Sears and Cushman signed on the dotted line for a mail-order distribution deal that must have expanded Cushman's market by tenfold. Sears called its scooters the Allstate, a fitting brand name as Sears saturated the U.S. with its catalog and outlet stores; throughout the 1950s and 1960s, Sears also imported Puch and Vespa "Cruisaire" scooters by the shipload.
Owner: Keith and Kim Weeks.

The Cushman Eagle

A Scooter for Real Men

In 1949, Cushman unveiled its new motorcycle in miniature, the Eagle. It was a radical departure from the step-through design to a scooter reminiscent of the big-twin motorcycles of the time that ruled the American roads—the Harley–Davidson Knucklehead and Indian Chief.

The Eagle laughed at the scooter design set in stone by Salsbury nearly two decades earlier. In creating the Eagle, Ammon and Herb Jesperson looked back to the Powell P-81 and the Mustang lines: The workings of the engine were proudly on display without engine covers and owners were not afraid to get their hands dirty working on these "real" machines. And these were machines for real men, riders who weren't afraid to swing their legs over the gas tank.

All in all, historical hindsight has crowned the Eagle as the right scooter at the right time for the U. S. The Eagle became the best-selling Cushman scooter ever.

▲ 1951 Cushman Eagle Ad
The first Model 765 Eagle Series I was built 1949–1954; in April 1952, the new square-barrel 5hp and the new 7.3hp engines were optional. From 1952-1954, the Model 762 Eagle was offered as a stripper economy model Eagle without the dreary hassles of a transmission.

> "Somebody in our sales department wanted a scooter that looked like a motorcycle with gas tank between your legs. It turned out to be a hell of a good idea."
> —Robert H. Ammon, 1992 interview

▲ 1965 Cushman Super Silver Eagle Series II
In early 1961, Cushman debuted an all-new, all-aluminum OMC Super Husky engine for the new Silver Eagle. But the new engine came about as a cost-saving effort by OMC; the first Silver Eagles literally shook themselves apart as they were held taut in the rigid Eagle frame. By 1962, the new Floating Power Chassis was developed with Power Frame rubber engine and transmission mounts to cope with a new Super Husky that included counterweights to balance vibrations. The Super Silver Eagle was truly "the big name in little wheels." Owner: Jim Kilau.

▸ 1960 Cushman Super Eagle and Eagle
While the rest of America was busy building bomb shelters, the smart set was out riding Cushman's great Eagle. The new Eagles combined "soaring ride with rugged roadability," according to this 1960 brochure. As always, Cushman ads were chock full of buzzwords to back the solid sales the Eagle line was turning in. "Just to look at the brand new Super Eagle and Eagle is to feel an urge to take to the road. Why not yield to that urge? Swing into the saddle, ease open the throttle, and learn a new definition of fun." Why not indeed?

◄ 1959 Cushman Super Eagle 765-88
The Model 765 Eagle Series II was created in 1955 with telescoping forks in the shadow of the Eagle's model, the Harley–Davidson big twin with its new Hydra-Glide forks. But the big change to the Eagle line came in 1959 when Cushman gave the Eagle a facelift with a "streamlined" rear body section. Owner: Roger McLaren.

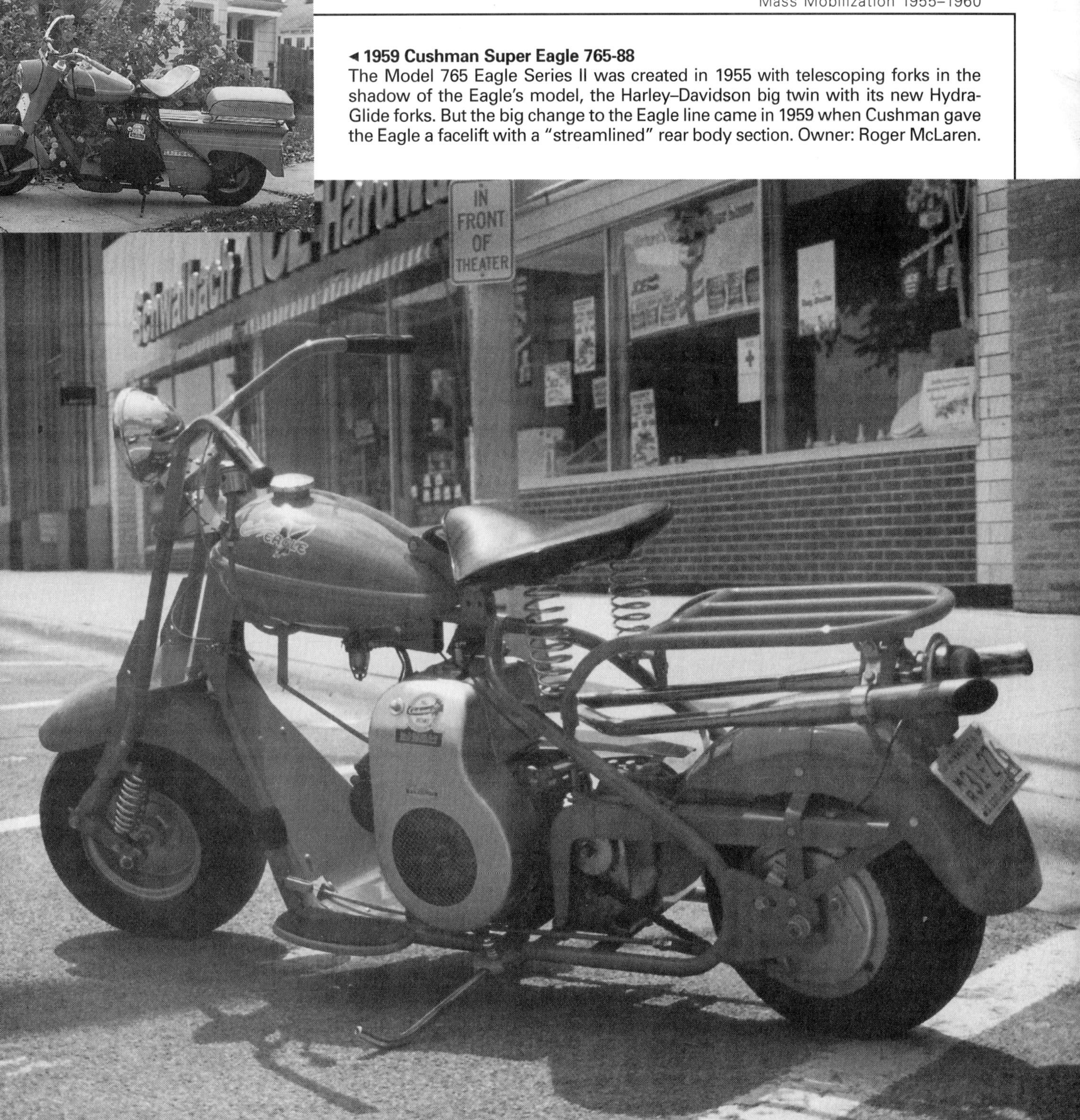

▲ 1951 Cushman Eagle
Just the thing to ride to the first of the new-fangled 3-D movies released in 1952, *Bwana Devil*, the novel Eagle took the mechanicals of the 60 Series step-through scooter and shoehorned them into a novel miniature Harley-Davidson chassis with a rigid rear—hardtail in Harley parlance. Bodywork was at a minimum to show off the Husky's muscle, as this was a scooter for real men. Debuting in December 1949, the Eagle was powered by the round-barrel 5hp Husky engine with a chrome-plated exhaust pipe swooping down to the right. The two-speed transmission was operated by a "suicide" hand-shift on the left side of the gas tank in the best Harley tradition. Drive on the first 765 Eagles was shifted from the left side to the right, the opposite of all preceding Cushman scooters. Owner: Curt Giese.

Aqua-Scoots and Sno-Scoots

No Water too Rough, No Snow too Deep

Once upon a time, Britannia ruled the seas. In the 1950s and 1960s, English entrepreneurs hoped to regain past glory by applying their seafaring knowledge to the ever-versatile scooter. These aqua-scoots, however bulky, inefficient, and unpopular, were the direct precursors of the Jet Ski, and equally as deafening to relaxing beach goers.

"Snow, however, is navigable if you know your stuff."
—*Popular Science*, 1957

In 1956, the Amanda Water Scooter lead the armada with the help of a Vincent engine, but its fiberglass hull melted down, drowning one test driver. Believing this mishap forgotten, a Lambretta conversion kit (applying ideas made popular in old *Popular Mechanics* magazines) used paddle blades attached to the rear wheels, which hopefully didn't spray too much water on the back of the first mate. This £25 kit appeared at the 1965 Brighton motorcycle show. As part of the ongoing Vespa-Lambretta rivalry, a Vespa was equipped with pontoons and crossed the English Channel in a historic publicity stunt.

▲ Land Ho!
Sure, Magellan traversed the globe by sea, but would he have been so brave as to traverse the English Channel by scooter? George Monneret boldly sailed from Calais to Dover on his 125cc Vespa outfitted with pontoons and paddles hooked up to the scooter engine. When he reached the white cliffs, he zipped up to London in record speed, making today's Chunnel seem positively sluggish.

► Scooter Party Barge
Even more proof that scooters can do anything, possibly even pull water-skiers. This 1939 aqua-scoot makes any tropical island a two-stroke paradise, at least until the tidal waves hit.

▲ **Snowmobile on Two Wheels**
With ad copy yelling the "Slimstyle by Lambretta is the absolute ultimate in motor scooters," these youngsters just couldn't resist hitting the slopes in their new Li. In true exclamatory duplicity, Innocenti bragged that the Slimstyle "handles like a bike, rides like a car." Some early pioneer motorscooters were even armed with treads on the bottom for maximum traction. And rumor has it that mad (garage) scientists in the frozen North even decked later scooters with skis and chains on their tires. Why they didn't just purchase a snowmobile is anyone's guess. *David Gaylin/Motor Cycle Days*

"I'm now working on a couple of collapsible pontoons and a propeller drive. So this summer, when I ride the scooter up to the lakeshore, I'll just keep on going. And maybe—souped up a little and equipped with rotor blades—it'll lift off the ground."
—A Do-It-Yourselfer quoted in *Popular Science*, 1961

Scooters and Sidecars

Bring the Whole Family!

Once pa married ma courtesy of their scooter-inspired courtship, they needed space for little junior on the putt-putt. Enter the sidecar, a one-wheeled, single-seat, bolt-on addition that allowed you to tote the whole family about like sardines in a motorized tin.

Sidecars were the craze in the 1950s and 1960s, a stepping stone on the way to a mini- or full-sized car for the growing family. Sidecars were built by makers around the world, and were attached to everything from Cushmans to Nortons.

But by the mid-1960s, sidecars were out of fashion as quickly as they had arrived. Mini-cars such as the Mini, Fiat 500, Volkswagen, and Citroën 2CV provided weather protection beyond simple legshields.

Call them a scooter with a bolt-on coffin or the precursor of the family safari wagon, that Heinkel *mit* Steib *seitenwagen* was once the rage of the roads.

► **1953 Goggo 200cc and Royal Sidecar**
The German Goggo scooter seemed perfectly matched to a sidecar, both sharing the same style of rounded fender and bodywork. It was not too great a leap of faith to simply enclose the whole kit'n'kaboodle within one sheet of steel and presto, a micro-car was born—which is exactly what Goggo did in creating its Goggomobil. The Goggo-Roller was available with your choice of 125cc, 150cc, or 200cc engine. The Goggomobil added a windshield, steering wheel, and funky suicide doors to the package.

▲ ***Motorroller mit Steib Roller-seitenwagen***
Steib of Nürnberg are famous the world over for their stylish sidecars. Steib built sidecars in a range of models from this scooter sidecar through sidecars sized for 250cc and 500cc motorcycles.

▲ **1949 Lambretta Model B and Sidecar**
Seeing the sights in a sidecar was certainly the way to travel, a modern mechanized version of the ancient Romans' horse-drawn chariots. This dapper duo toured the piazzas of Milan in style. *Collezione Vittorio Tessera*

▲ 1947 Salsbury Model 85 and Sidecar
The aliens have landed! Driving down Sunset Boulevard in this Salsbury duo would have evoked cries of fear that Orson Welles' *War of the Worlds* radio show had actually come true. Unfortunately, this Salsbury "prototype" existed only on paper, an artist's airbrushed conception of what such a scooter spaceship could have been. Alas. *E. Foster Salsbury Archives*

Golf Carts: Scooters for Grown-Ups

Just as children's push scooters begat motorscooters, motorscooters begat golf carts. Those grown-up kids in kelly green who once buzzed Main Street on a Cushman Auto-Glide could now tour the links in a Cushman Golfster golf cart. Out for some sun, fun, and eighteen holes, some golfers even putted from their putt-putts. In the US of A, both Cushman and Harley-Davidson stopped building scooters and tooled up for golf carts. It was the dawn of a new Golden Age for scooter makers in America and Europe when the scoot market went soft and flabby around the middle.

► Grand-Duc of Luxembourg and EZGO Golf Cart
The tiny grand duchy of Luxembourg's Grand-Duc, Grand-Duchesse, and heirs proudly pose with the family EZGO golf cart.

Micro-Cars, Mini-Cars, and Bubble Cars

The Motorscooter's Big Brother

▲ 1958-1961 Piaggio Vespa 400
It was a logical step from the Vespa motorscooter to the Vespa micro-car for Piaggio. The Vespa 400 was unveiled in 1958 with a front-mounted two-cylinder 394cc engine capable of pumping out 14hp.

Micro-cars were motorscooters that ate their Wheaties and grew up. Where push scooters were fitted with engines to beget motorscooters, in the case of micro-cars it seemed as though someone somewhere had taken a notion to try the same thing with junior's pedal car.

Micro-cars were just that: Microscopic-sized cars, at least in relation to real cars, such as Volkswagen Bugs. They were often powered by scooter or motorcycle engines, fitted with a roof, door(s), full seats, and more than two wheels, although not always four. Some even had steering wheels instead of handlebars. Micro-cars were uptown.

Micro-cars played a role in Europe's reconstruction just as the scooter did. Alas, even today Isettas, Fiat 500s, and Minis are simply ignored, made jokes of, or tripped over. Except by the faithful, of course.

▲ Isetta Invasion
Scooter maker Iso crafted the design for the Isetta based on inspiration—no joking—from a watermelon. Iso didn't have the finances to build many Isettas, but it licensed production of the three-, and later four-, wheeler throughout the world and used the royalties to jump from scooter production to building *gran turismo* automobiles to rival Ferrari and Lamborghini. Isettas were built by Hoffman and BMW in Germany as well as in Spain and Brazil; BMW built thousands of Isettas propelled by its one- and two-cylinder motorcycle engines.

◄ British Ad for Scooters and Micro-Cars
This ad from *Motor Cycling* in 1957 showcases one dealer's wares ranging from scooters to three-wheeled micro-cars—all available "in a wide range of modern colours." Micro-cars were not a large step upward from scooters; they were typically based on scooter or motorcycle engines, added a third wheel for stability, an extra seat or two, and an enclosed top and doors—usually. The AC Petite, Reliant Regal, and scooter-maker Bond's Minicars were typical of the early British mini-cars; the next generation of mini-cars was heralded by the four-wheeled arrival of the well-named Mini, as well as Italy's Fiat 500, Germany's Isetta, and France Citroën 2CV.

◄ Messerschmitt KR200
Shifting from Luftwaffe airplane production during World War II, Messerschmitt license-built Vespas for the fatherland before creating in 1953 its superlative bubble cars. These three- and four-wheelers were the coolest cars on the *strasse*, wowing the *volk* around the world with their miniature technology even today.

CHAPTER 6

The MOD YEARS

1960-1975

One of the most enduring styles that sprang from the back streets of London is that of the Moderns. In the early 1960s, the standard of living was on the rise. While mum and dad picked up an auto, the new teenager market demanded a high-performance steed for their friends to ogle while they zipped by the coffee shop. These scooters, however, came as an afterthought to the Mod agenda: another piece of jewelry to match the slick Italian clothes and the mop-top dos.

The Mods were preceded by the Teddy Boys and rivaled by the Rockers. The dapper Teds outdid each other in their slick outfits and patent leather shoes as they bopped to the music of American rock'n'roll. The Rockers, on the other hand, slicked their hair, wore leather jackets, and weren't afraid to bop a passing Mod. In general, however, Rockers didn't pay the Mods much attention since they were busy popping coins in the Ace Bar jukebox, kick-starting their Café Racers, and blasting around the block before the song ended. Unphased by the competition, Mods sat proudly on their Lambrettas, or Italian hair dryers as the Rockers called them.

Mods' scooters went through many phases as they constantly tried to outdo each other. In the late 1950s, the early post-Teddy Boy phase caught everyone listening to cats like Monk and Mingus and driving their stock scooters fresh from the crates. Two-tone paint jobs soon became the rage. The mid-1960s saw decked-out Lambrettas with every bell, light, and horn they could get their hands on and blasting the Who and Otis Redding.

In 1964, the Mod-Rocker wars climaxed with a violent melee on the beaches of Brighton. In a promotional frenzy, Innocenti sponsored a scooter meet to the delight of Mods from all over Britain. The event fell on a bank holiday, so numerous Rockers had road-tripped to the sea to stroll along the boardwalks and blast some four-stroke exhaust in the faces of their two-stroke rivals. Soon, scooters were tipped, beach umbrellas thrown, and hairdos mussed. The land of tabloids was willingly shocked as the exaggerated rumble was splashed on the front page of every newspaper.

Mods began as mostly well-off West-Enders who hung out on Carnaby Street in search of some fun. They weren't afraid of ruffling a few feathers with greasy rockers on their BSA Gold Stars and Triumph Bonnies just as long as their Lammies would get them to the club in time for the Skatalites' show.

"The Mod way of life consisted of total devotion to looking and being 'cool.' Spending practically all your money on clothes and all your after-hours in clubs and dance halls."
–Richard Barnes from *Mods*

▸ **Mod Memorabilia**
Riding Vespas and Lammies fitted with enough headlamps to light the night and wearing skinny ties, army parkas, and desert boots, the Mods were modern to the hilt. Gee-tar of choice for the Mod Pete Townshend and his R&B band The Who was the Rickenbacker; this is a U.S. Model 360.

ROCK
Mods!
JACK KEROUAC
THE DHARMA BUMS
VIKING
55 6 7 8 10 14 16
MOTOROLA
THE
PERSONALISE
with genuine
accessories
Lambretta

It's a Mod, Mod World

Their God Wore Shades

The world was indifferent to Mod scooters–referring to them mockingly as "Italian hairdryers"—until they decked them out with every headlight, mirror, horn, and any other chromed piece of equipment they could put their hands on. Now, they couldn't be ignored as their scooter jewelry on their weighted-down steeds blinded other drivers and pedestrians who shielded their eyes to the reflecting sun from their chrome trinkets. The Mods, however, merely accentuated a trend to gussy-up their scooters that began in the 1950s and later continued with customized Harley hogs. Anybody who was anybody had a scooter.

"I don't wanna be the same as everybody else. That's why I'm a Mod, see?"
—Jimmy, *Quadrophenia*

▲ Modfest
Get out your skinny ties and pointy shoes, fire up your two-tone two-stroke and hit the dance floor. Skafests still happen at least twice a year in most major cities all over England and the U.S. Ploys like "Last gig ever?" by the International Beat, once the English Beat then Special Beat, are typical promotional fare making this once again the show *not* to miss.

"You've got to be either a Mod or Rocker to mean anything. Mods are neat and clean. Rockers look like Elvis Presley, only worse."
—Mod Teresa Gordon quoted in the *Daily Mirror*, 1964

▲ "Rumble in Brighton Tonight!"
So sang Brian Setzer of the Stray Cats more than a decade after the infamous Mod-Rocker battles on the beach of Brighton. This still from *Quadrophenia* shows the Rockers on their Triumphs and BSAs taunting the Mods in their hooded parkas and dapper outfits. Even though the Who's movie came after most of the rivalry between the Mods and the Rockers had calmed down, *Quadrophenia* still remains the fictional documentation of the era inspiring many a new Mod to fix up an abandoned Lambretta.

▲ **Lambretta Jewelry**
Realizing that scooter owners often spent as much money on accessories as they did on the actual vehicle, companies like Viganó Mettallurgic near Lake Como offered their own chrome products. This Lambretta jewelry often went as far as clandestinely copying the actual name of the scooter and model into the chrome.

▲ **Italo-Mods**
In a bizarre twist, these modern Italian Mods copy their British counterparts who originally copied Italian style. And to be truly Mod, they have to assume stylish English names, like Senior and Oscar, and play in a band. Poor Rudy might die of humiliation; he has the quintessential Rude Boy name but no scooter. Oh, for shame.

▲ **Might as Well Carry an Advertising Banner for Lambretta.**
Bizarre aftermarket accessories were made by a number of firms while the genuine Lambretta ones often lacked the panache of smaller companies. Mods wouldn't settle for the ordinary, so they raided hardware stores for any lights, flags, or mirrors they could throw on their steeds. Although many scooters were run on magnetos, all the extra lights required battery systems to be installed.

1993 Mod Revival

"What is a Mod?

"Mod, Modern. It means now, here, us. It means motion. Moving towards the future. It means accepting the trash, the cool, the end of our world. *In fact it means bringing it about ourselves*. With style. The fasions [sic] and fetishes of our world. The black and white clothes, the black and white kids. All of us throwing switches and watching electricity power our motors. Looking up at the buildings that spiral up into the sky. This is our world, this is our time. Mod means living today, sensing tomorrow, remembering yesterday and all it's [sic] pain and love. Mod means us. It means self definition. It means unity and individuality. It means never having to grow up. It means the rituals of modern life—the schoolyards, subway station, the diners and coastal towns. The parties where nobody ever, ever has fun. It's not a wonderful life. It's just a life, the only one we have, and we're not going to live a rerun. We don't need their politics, their ideas or their heroes. We have their technology, and we have our formula."

–excerpt from *Absolute Beginners* magazine, 1993

Rock on the Roll

Hymns to the Motorscooter

"Get on your bad motor scooter and ride/When the sun comes up, everything gonna be all right."
—Montrose,
"Bad Motor Scooter"

Scooters have probably inspired more innovative swear words than heavenly hymns. But there have been several songs sung to the scooter.

In 1947, Innocenti announced the arrival of its first Lambretta scooter with a catchy advertising tune played night and day over RAI radio. The ditty was one of those "Shave-and-a-haircut-two-bits" opuses that you could not get out of your mind no matter how hard you tried—and is still whistled to this day by Milanese caught off guard.

Then in the 1960s came the Italian hit, "Lambretta Twist," and on dance floors everywhere the crowd was twisting—everyone, that is, except Vespisti.

The most eloquent testimonial to the scooter was surely The Who's *Quadrophenia*, a rock'n'roll opera telling the tale of the Mod-Rocker clashes in 1960s England. A Vespa graced the album cover as well as the movie poster, and the sinking of a scooter ended the film with symbolic eloquence. Throughout, The Who were turned to eleven with Pete Townshend smashing Rickenbackers through Hiwatt stacks and Roger Daltry yodeling in "The Punk Meets the Godfather": "I ride a GS scooter with my hair so neat/Wear a war-torn parka in the wind and sleet."

▲ **Spinning Disks with a Zündapp Bella**
Roll over, Wagner. A bucolic idyll—about to be shattered forever by the twang and fever of some good oompah music blasted out of a windup record player at maximum distortion. *Collezione Vittorio Tessera*

▲ Rocking Scoots
Two of the rockingest scooter albums ever. The Who's *Quadrophenia* was a rock'n'roll ode to the Mod-Rocker wars of the 1960s as well as to the Vespas and Lammies the Mods rode. In the flick, Sting makes his appearance as the quintessential Modern, complete with scoot. Bo Diddley, meanwhile, had guitar and traveled on his Allstate Jetsweep. This 1960s Bo album did not feature any references to his motorized friend, but Bo's signature riff may have been inspired by the beat of his scooter's heart. Both albums should be played loud and proud. The gee-tar is a rockabilly-approved 1956 Gibson ES-295 leaning against a 1964 Vox AC30.

The 1970s brought us glam rock, afros, platform boots, and the quintessential seventies band Montrose with its hard-rocking hit, "Bad Motor Scooter." Complete with revving-engine guitar sounds (probably made by the rare and coveted Hemi stomp box), this was the scooter's *Easy Rider* anthem destined to set putt-puttniks' hearts aflutter: "Get on your bad motor scooter and ride/When the sun comes up, everything gonna be all right."

Amen.

"I ride a GS scooter with my hair so neat/Wear a war-torn parka in the wind and sleet."
—The Who, "The Punk Meets the Godfather"

Movie Stars and Motorscooters

The Putt-Putt on TV and Film

While General Motors had Pat Boone as their mascot, scooter makers couldn't afford the big guns, so they settled for paparazzi photos of stars on their putt-putts. Perhaps the most famous scooter photo of all time is Amelia Earhart on a Motoped which did more for promoting motorscooters than any number of clever ad men. In a major coup for Innocenti, Paul Newman appeared on the cover of *Scooter* magazine piloting his Lambretta through heavy traffic. And guitar hero Bo Diddley featured an Allstate Jetsweep scooter on the cover of his album complete with a square-body Gretsch guitar.

Scooters in film, however, gave them more hipness that any individual photo could possibly achieve. *Roman Holiday* with Audrey Hepburn and Gregory Peck showed them zooming around the Eternal City in scenes that to this day encourage tourists to pay huge fees to rent Vespas. *Quadrophenia* showed primped mods on Lambrettas battling the naughty rockers on BSAs to the blasting sound of Pete Townshend's wailing Rickenbacker.

Federico Fellini's *La Dolce Vita* featured newsman Marcello Mastroianni in his Lancia spyder while whirring Vespas encircle his car in hopes of getting the scoop first. Mastroianni always has an angle on news stories, but could never quite escape from the flock of scooters that constantly swarm him wherever he goes. The most famous of these muckrakers was the photographer, Paparazzo, whose scooter transported him to remote hideaways of the stars, so he could snap scandalous photos.

▲ **Kookie's Topper**
Everyone's favorite star of *77 Sunset Strip* takes advantage of a photo op to appear on Harley Davidson's only scooter, the Topper. It's unknown if Ed "Kookie" Byrnes' career would have been more successful if he hadn't appeared in this compromising snapshot, but at least he had the opportunity to handle the "perfect roadability" made possible by the Scootaway automatic transmission.
David Gaylin/Motor Cycle Days

> "Everyone knows that damage is done to the soul by bad motion pictures."
> —Pope Pius XI

◄ **Burl Ives Goes Scottish**
Burl Ives shares a laugh with a pint-sized admirer on a Lambretta LD. The singer was usually spotted with his guitar and legions of fans, so this snapshot was definitely an Innocenti publicity coup to inspire folkies everywhere to go two-stroke. One of the original design features of scooters was the open area between the driver's legs to allow dresses, or kilts, to be worn.
David Gaylin/Motor Cycle Days

► **Spanky's Safety Patrol**
The little rascal becomes a little cop as he zooms these youngsters to safety in his Powell Streamliner scooter with sidecar. Although the scowling youth was premature to drive a scooter, Powell's ads boasted anyway, "Everyone who can ride a bicycle can ride a Powell Motor Scooter!"
Herb Singe Archives

STOP
SPANKY'S
SAFETY
PATROL

How to Drive Like an Italian

Ride Loud, Ride Fast

There's only one true rule to driving in Italy that all must follow: It is illegal to allow someone to be in front of you. All traffic tricks, go-fast fantasies, two-wheeled cornering maneuvers, and one-wheeled braking dance steps derive from this basic maxim.

So whether you are driving a Lamborghini Miura or a Lambretta LD 125, the rules of the road are the same—except with a motorscooter you have more to prove.

▲ Loud Fast Rules
Drive fast, drive loud is the motto for scooterists in fair Italy where every stoplight is the starting line of a grand prix and every ride is a race. Scooters were not made to stroll; they were built to zoom. In the Stoplight Grand Prix, the power-to-weight ratio of a Vespa allows you to beat out a Ferrari for the first 10ft, so always take advantage of this to blow some two-stroke smog in the *gran turismo* snob's snout. Every Ferrari and Fiat will then pass you up the *strada*—until the next stoplight, when the scooters sift back through the traffic to the front, and puff off at the starting line once again to repeat the whole process. If you must drive slow, be certain you're in low gear so the engine roars at high rpm. The sound will ricochet off the walls of the houses lining the streets, waking everyone from their siesta. Don't forget to saw off a good 6in of your exhaust pipe for maximum volume and speed. Just remember: Loud fast rules! *Collezione Vittorio Tessera*

▲ Accessorize for Speed
As in the fashion world, accessorizing your look is the key to dressing for success. With scooters and their puny top speeds, you can't always drive fast, thus you must at least look like you're driving at maximum velocity. This 1950s Vespa is decked out for piazza cruising with so many aftermarket accessories that it attracts *carabinieri* to write up a speeding ticket even while it's parked: Chrome trinkets, shiny but nonfunctional engine trim, big-lipped bumper overriders, wire headlamp stoneguards, whitewall Pirelli tires, spare wheel with leatherette cover, metallic cable covers, and a St. Christopher's medal are almost essential as a starting point. In the land of bolt-on chrome exhaust extenders, you can never accessorize too much. *Collezione Vittorio Tessera*

▲ Hand Gestures Are Required
Often mistaken for a photo of an Innocenti stunt team, this is in truth a family piled onto a Lambretta LC 125 who have just been cut off in traffic and are responding with the required hand gestures. In the early days of scooting, whole households rode to market or the beach on the family putt-putt, even though the two-stroke engine strained to haul but a single rider. Hand gestures for left and right turns are optional, but imaginative gestures are *de rigeur* especially when you're snubbed in traffic. Even more important, however, is the need to interact with your passengers, even to the point of ignoring traffic. The hospitality extended to your passenger(s)—especially those of the opposite sex—requires a rare dexterity of hand gestures to accentuate a conversation. In addition to words and hand gestures, the art of Italian conversation also calls for eye contact at all times, a true trick as you navigate through laneless streets, gesticulate to a Ferrari driver to help him improve his driving habits, and flirt with your pillion compatriot.

▲ *La Bella Donna* and *La Bella Figura*
The object of riding a scooter is maximum visibility to create *una bella figura*, or "a beautiful figure"—better known in American slang as simply "being cool." The trick, however, is to never, ever act as if you realize you're making a scene—that would be creating *una brutta figura*, "an ugly figure." There are several tricks to the trade here. First, if your Lambretta is coughing out vast billowing clouds of exhaust, remedy this tuning oversight by wearing Ray-Bans. Second, if your engine seizes at speed, simply broadslide the scooter to a stop and pretend you were going to park or had planned on ditching the thing right there anyway. Among the worst moves you could make is having your scooter break where the coffeeshop crowd can see you. The rule here is to never repair a broken-down Vespa on the side of the road: Not only will you be visible to the passersby, but you also stand a chance of getting oil on your Versace clothes. A scooter breakdown can only happen while riding with a member of the opposite sex and can only be intentional, usually occurring in a secluded, wooded spot.

Toys-R-Scoots

Putt-Putt Playthings

▲ **Vespa Toy**
Most motorcyclists already consider Vespas to be toys, but this toddler's scooter showed that *real* toys have training wheels and come in plastic. Unfortunately, lack of a battery meant that mini-Mod wannabes couldn't attach every halogen bulb they could get their hands on. Owner: Herb Singe

The earliest scooters developed from the idea of children's push toys with a motor attached to become a viable and efficient form of transport. Then in the 1950s, when scooters buzzed through fashionable avenues the world over, children's toy manufacturers once again turned to the push scooter, only this time reproducing the exotic European designs.

Motorscooters inspired mini scoot-like creations for tots to emulate their elders. While many considered full-sized scooters to be toys already, at least they didn't offer training wheels like some of their pint-sized half brothers.

◄ **Dueling Scooter Toys by Technofix, 1950**
These handpainted West German toys feature two inseparable Siamese twins. These two scooters come from identical metal molds, but are each painted differently with superb attention to detail and training wheels to boot.

"To turn a fourteen-year-old child loose on a motorscooter in today's traffic is about as sensible as giving a baby a dynamite cap for a teething ring."
—Gen. George S. Stewart, National Safety Council, 1959

◂ Russian Moto Roller
While the Viatka stole the design of Piaggio's Vespa behind the patent-free safety of the iron curtain, Soviet toy companies took it a step further and copied the lines for their toys. With even smaller wheels and a more upright front fork than the motorized version, the Moto Roller when driven on cobblestone streets would sure make Junior want to move up to the real thing.

◂ Starter Scooter
Just what every child dreams of receiving on Christmas morning, and will be sure to break before New Year's Day. With a dragging front kick-stand that is certain to amuse friends with sparks, this starter scooter was a hit at the famous Blind Lizard Run in Minneapolis.

Microscopic-Sized Mini-Scooters

Voodoo Putt-Putt (Slight Return)

The next step in the mobilization of the masses lay in mini-scoots. Just pop them in the trunk of your car and where the road stops, the fun begins. No more of that tiresome walking to get from point A to point B with these petite putt-putts; you can drive right to the door, fold up the scoot, and carry it with you—all 50lbs.

Although they may appear to be condensed versions of larger scooters, these mini powerhouses often contained potent little engines with a good power-to-weight ratio on rare aluminum frames that zipped the rider along with no protection whatsoever. Luckily, the speeds made getting hurt almost impossibile.

The target markets for mini fold-up scooters were boaters, pilots, and especially folks in starter-homes at trailer parks. In raving over the Centaur scooter, one ad copy writer crowed, it's "a perfect blessing around the trailer park or campsite and folds away so neatly when we're on the go."

Carries this 210 lb., 6' Man, 15 M.P.H., with ease.

MOHS ELECTRIC SCOOTER

For Industrial, Institutional and Pleasure Use

- **QUALITY CONSTRUCTION** — High strength 24 ga. paint-lock galvanized body mounted on arc welded steel frame of unique design. Body and frame finished in durable metallic lacquer or enamel. Heavy duty Goodyear tires and roller bearing wheels. Long lasting golf-cart type battery made for complete discharge and rapid recharge. Sturdy electric motor running at half of designed voltage for long-life driving rear wheel with Diamond No. 35 chain — no power-robbing jack-shaft required. Internal expanding brake actuated by hand-grip on right handlebar. Power switch on right foot. Both easily altered to accommodate the handicapped. Upholstered in Premium Naugahyde, the finest and thickest available over foam cushion. Can be finished in your firm's colors at slight extra cost.
- **USEFUL** — Ideal for plant personnel use and inspection trips. No noxious fumes or excessive noise make the Mohs Scooter particularly useful indoors. Efficient in all temperature and climactic extremes. Starts instantly always for the man with too-few hours in his day.
- **PERFORMANCE** — Carries the 210 lb., six-footer illustrated in photo above with ease over eight miles on a single battery charge at speeds up to 15 M.P.H. on constant discharge. Correspondingly greater distances, up to 16 miles, with a lighter man on board and intermittent discharge. Battery recovers quickly during lunch hours or overnight simply by plugging built-in charger into any 110 volt, A.C. outlet. Easy to balance and highly maneuverable. Turns in 4' radius circle under power without backing.
- **TO ORDER** — Tell us your intended use, color desired and any special painting instructions, trade-mark decals or any other specials you wish installed, your complete address and preferred mode of shipment. Price effective Oct. 1, 1963, F.O.B. Madison, Wis.

$225.00 Standard.
235.00 As above with special paint or decals to your specifications.

Customs duties on foreign delivery to be paid by purchaser. Price includes Wisconsin and Federal taxes.

Mohs Seaplane Corporation

▲ Mohs Electric Scooter
The Mohs Electric Scooter was so ahead of its time that years passed before anyone could see the idea of electric scooters as anything more than a liability. In this Darwinian world, however, when the use of something isn't immediately appreciated, it ends up on trash heaps. Not until the early 1990s did large scooter companies begin to consider the advantages of clean electric fuel. The selling gimmick of this particular scooter was its ability to carry heavy people since its other qualities, a maximum speed of 15mph and a range of 16 miles per trip, left something to be desired. *Herb Singe Archives*

▲ 1946 Beam Doodle Bug
To many, the Doodle Bug is synonymous with scooter. Marketed throughout the country by Gambles Stores, this putt-putt was every American youth's dream. The simple design of Beam's scooter is just one step up from a child's push scooter, making it the perfect gadabout for junior while dad is cruising on his Harley. Owners: Keith and Kim Weeks.

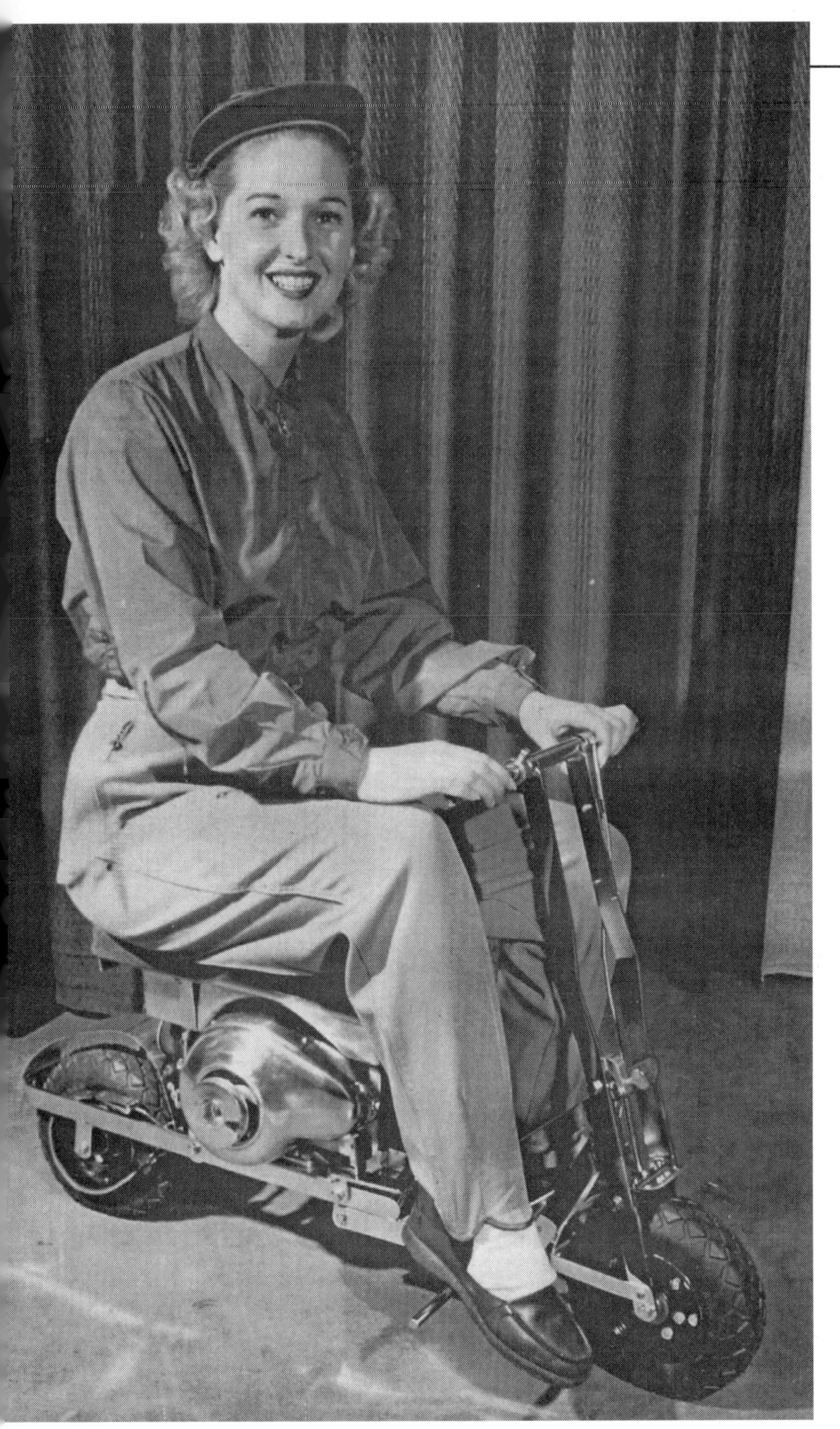

▲ 1950 Argyle Scooter Cub and Happy Scooterist
Amidst motorboats and Winnebagos, lovely Miss Rita Barry shows off an Argyle scooter at the Chicago Outdoors Show in 1950. Once again, the highlight of this Scooter Cub is that it "Will carry two heavy people," and if that's not enough, it folds up like a briefcase–a 50lb briefcase–in just 15 seconds and fits in your car, boat, or plane. Try convincing customs agents that this mysterious gas-filled, metallic object isn't for a reenactment of the Hindenburg. At shows like this, the Argyle attempted to conquer an untapped market, the mobile-home bunch, since "trailer enthusiasts particularly like the convenience of the Scooter Cub." *Herb Singe Archives*

▲ Skat-Kitty and Sidecar
Projects Unlimited of Dayton, Ohio, produced the miniature Skat-Kitty as a powerful, functional scooter. Although the logo of the Skat-Kitty showed a goggled kitten zooming along at breakneck speed, it's doubtful if the scooter ever broke the 20mph sound barrier. Owner: Herb Singe.

▲ 1950s Gresvig Folkescooter
This miniature marvel was built in Norway in the early 1950s and in Sweden as the Mustad; only the names were changed while the mechanicals remained the same. Both scoots were powered by 49cc two-stroke engines pumping iron for 0.8hp. The front-mounted engine drove the front wheel via a chain, marking this "People's Scooter" as one of the few front-wheel-drive putt-putts. *Ole Birger Gjevre*

Motorscooter Believe It or Not!

This Really Happened—Really!

Not only is truth truer than fiction, it also tends to be more outrageous. This age-old maxim is especially true when it comes to motorscooters.

Believe it or not, the scenes you are about to see actually did happen. The proof is in the pictures.

"In Spain, there has emerged a new style comic art. A garishly costumed toreador 'fights' the bull on a Vespa between acts of an orthodox bullfight."
—*American Mercury*, 1957

▲ Motorscooter Toreadors
Bullfighting on motorscooters was like playing Bizet's *Carmen* on electric guitar. Sacrilege? Perhaps, but both have been done. These mechanized toreadors straddle a Parilla Levriere scooter as they prepare to lance a bull before the corrida crowd in Lisbon, Portugal, in 1956. No doubt the scooter was painted bright Ferrari red to lure the bull as the riders crossed their collective fingers that the engine did not die at an inopportune moment. This really did happen.

▲ Scooters on Ice
The 1955 IceCapades show toured the United States featuring such spectacles as a re-enactment of Peter Pan on skates as well as the Fantasy in Pink grand finale of Ravel's Bolero. The true spectacle, however, was Autorama: America's fascination with the motorscooter came alive in song and dance, choreographed on ice with no less than 19 Lambretta LD 125 Series 1 scooters. The traffic police rode their putt-putts on ice without augering in, rivaling even the Shriners in scootering prowess. Meanwhile, the Ice "Cop" Ades, Ice Ca "Pets," and the El-Dorables pirouetted around the buzzing scoots wearing gas pumps on their heads as well as skating outfits with car grilles on the front and hood ornaments on their tiaras—all while trying not to choke on two-stroke fumes. This, too, really did happen. *Hymie's Vintage Record City*

Motorscooter Heraldry

Stars, Arrows, and Coats-of-Arms

Even if you never see the scooter, chances are you've seen the design of the emblem. Millions of copies of the logo and font used for the name of the scooter and its company were reproduced on everything from brochures to posters to letterhead to the actual scooter emblem. Nearly every molded metal stock part and accessory contained the logo of the parent company or the specific model. What's the point of parking a Lambretta TV Series 2 in front of the Café di Roma if gawkers can't tell immediately what brand of scooter you're driving?

▲ Vespa Symbolism
The look of the Vespa logo changed with the years. The original Piaggio logo of 1946 in the upper right corner was mass produced in a metal base with baked enamel and appeared on the front horn cover of the older Vespas. The cursive Vespa logo, which was attached to the legshield, is representative of the attention given to font styling in Italy in the early part of this century. The upwardly slanting diagonal line underneath the lettering means motion and change as opposed to the conservative Vespa lettering of the seventies in the lower right corner.

Shriner Motor Corps

The Right Stuff

Shriners are to motorscooters what Chuck Yeager and his band of test pilots are to super-mach jet fighters. When it comes to scooters, the fez-capped Shriners have The Right Stuff.

Who hasn't been stunned by a Shriner Motor Corps' dazzling riding skills, intricate death-defying figure-eight maneuvers, and devil-may-care roundabouts at a Fourth of July Main Street parade—too stunned to even move your hand to catch the taffy the Shriner jocks deftly toss to you at the finale. Liken them to the US Navy's Blue Angels or the Air Force's Thunderbirds if you will, Shriners and motorscooters were made for each other.

▲ **Shriner Memorabilia**

▲ **Shriners on Parade**
Fast as speeding bullets, these Shriners from Washington state's Columbia Temple perform a dazzling display of scootering prowess. It's all in a day's work, of course. For this Fourth of July parade in Long Beach, Washington, they ride an assemblage of early Cushman Eagles and late Super Silver Eagles, all decked out with special options that mere mortals cannot buy. And the crowd is clapping hard and furious. *Courtesy Noble Rick Watt*

► **Shriner with Cushman Super Eagle**
Hailing from Spokane's El-Katif Temple, here is the well-equipped Shriner Motor Patrol rider with his trusty Super Eagle about to embark on a Fourth of July parade display. From Cushman's special Shriner Option Package, this scooter is decked out with saddlebags, flag holder, windshield, oompah air horn, and chrome on most everything that can be chromed. *Courtesy Noble Frank Workman*

▲ Hejaz Motor Corps and Cushman Eagles
Sponsored by the Yazoo Lawn Mower Center, an authorized Cushman dealer, the Hejaz Temple Motor Corps was ready to wow the crowd at a Fourth of July parade in their hometown of Greenville, South Carolina. Thirty riders strong, their figure-eight maneuvers must have been a sight to see back in the 1960s. *Photo courtesy Shriner Sam Nelson*

The Shriners' Trusty Steeds

All the Pretty Motorscooters

The Shriner fad for scooters started with Cushman chief Bob Ammon's brother, Bill. Bill was a long-standing Shriner, and offered Cushmans bedecked with accessories and chrome trinkets to his brethren at bargain prices.

It's fitting that these "Turtles"—to appropriate a jet test pilot slang term for the best of the best—first rode Cushman Turtlebacks. Just as a Harley is the motorcycle of choice for the Hell's Angels, the Cushman is the preferred ride for most Shriners. Sure, you may have witnessed their prowess on a Vespa, Honda, or Harley–Davidson Topper, but Shriners and Cushmans go together like Shriners and fezzes.

▲ **Shriner Motor Patrol and Cushman Super Eagle**
White scooter, whitewall tires, and white driving gloves: this Shriner Motor Patrol rider from the El-Katif Temple in Spokane, Washington, is class on two wheels. *Courtesy Noble Frank Workman*

▲ Shriner Spit and Polish
The Motor Patrol from the El-Katif Temple in Spokane, Washington, shows off their line-up of Cushman Super Eagles decked out in a dazzling array of special Shriner options. Cushman alone offered these accessories to Shriners on its scooters as well as the Vespas. Cushman began importing in 1961; only later did other scooter and motorcycle makers, such as Harley-Davidson or the American importers of BMW and Japanese machines, jump on the bandwagon with Shriner accessories. *Courtesy Noble Frank Workman*

Cushman's 1957 Shriner Option Package

Beginning in the mid-1950s, Cushman catered to Shriners with an arm's-length list of optional goodies:

- Buco black-and-white saddlebags complete with fringe and rivets
- Buco fender flap with rivets and reflectors
- Shriner emblems on most surfaces of more than 4sq-in
- Speedometer
- Windshield
- Multiple horns
- American and Shriner flags with whip-style flagpoles
- Super special paint jobs with lot prices
- Chromed safety bars, fender tips, gas tank bands, locking gas caps, floorboards, stands, clutch arms, air shrouds, front seat rails, and rear seats with chromed seat rails

◄ Shriner Motor Patrol Drill
The eight 1963 and 1964 Cushman Super Silver Eagles of the Lincoln, Nebraska, Sesostikis Temple Motor Patrol are almost identical in every feature. For judging competition with other temples, they are maintained in immaculate condition even today, thirty years after they performed their first pirouettes and figure-eights. *Courtesy Sesostikis Motor Patrol Captain Jack Douglass*

Chapter 7

BACK TO THE FUTURE

1976-On

The United States, Europe, and Japan gave birth to motorscooters, but the economic climate has progressed so dramatically that the days of scooters buzzing down main street have gone the way of the horse and buggy. Nostalgia runs rampant for when Vespas, Fujis, and Cushmans reigned supreme in the town piazzas and pedestrians complained of scooters' pollution, noise, and dangerous speeds rather than that of automobiles.

The good ol' days, however, live on in the form of retro scooter clubs idolizing the past and digging old Lambrettas out of barns and fixing them up better than new. Driving a restored scooter down the road now turns heads like never before, and old-timers aren't afraid to tell tall tales of bygone days of Cushman splendor.

More significantly than clubs and restoring the old is the Third World boom with more scooters than any of the developed nations ever saw zooming their streets. In fact, more scooters are in use today than ever before. Honda has produced more than one million Cub scooters that carry drivers from Mexico to Sri Lanka, and Yamaha and Suzuki are close competition.

The largest producer of scooters in the world, however, lies in India. Called Bajaj, it makes Vespas like never before. Not to be outdone, however, Slimline Lambrettas are still being produced in vast numbers by Scooters India with ever-improving technology. Old rivalries never die as the Vespa-Lambretta wars live on in a different part of the world. Although tough environmental laws make these scooters hard to obtain in the Western world, they are often available as some-assembly-required kits.

The worldwide scooter boom may seem like it's over, but it has just switched theaters and shows no sign of subsiding.

"Motorscooters have always been Warm Puppy Good Things, symbols of happiness and frivolity in days when motorcycles were symbols of malevolence, darkness, and perversity."
—*Cycle*, September 1981

▸ **Scooting into the 1990s**
Scoots are still alive and well in the 1990s and looking forward to riding into the next century in style. Whether it's the latest Piaggio, Honda, or Yamaha, or a back-to-the-past scooter revival on restored Cushmans or Vespas, the faithful still ride on two wheels.

HELMETS
YAMAHA
ELITE
SONY

Into The 21st Century

To Boldly Go Where No Scooter Has Gone Before

Scooters have inspired many great minds to make brilliant, but often ridiculous, inventions, and the tide shows no sign of subsiding. Honda makes its mini-Gold Wing scooter, the Helix. Unique Mobility is working on a modern electric scooter. And Piaggio continues to design ever more bizarre-looking Vespa clones. The trend shows no sign of an end.

After the 1960s, scooterists were often looked on with pity as poor folk who couldn't afford a real car, such as a Mini. Scooters were viewed as a passing fad that had had its day. The final nails in the coffin were worldwide environmental laws putting restrictions and heavy tariffs on two-stroke engines. New, economical engines were developed to abide by these laws, and when these new scooters couldn't meet the restrictions, they moved to the Third World. The scooter remains undead, however, with renewed interest by new generations who want an economical and convenient form of transportation that will zip them away from the clogged interstate to greener pastures.

▼65mph Barca-Lounger.
The Honda Helix was a sensation when it hit the streets. Like an E-Z Boy on wheels, the four-stroke, liquid-cooled Helix was adored by owners and ridiculed by spectators. No matter how you looked at it, the 250cc Helix was one of the fastest scooters on the road, and could easily zoom onto the freeways of the world without any problem, except for distracting other drivers who were amazed at a 65mph Barca-Lounger.

▶ Talking Scooters
While the market for scooters in the West has declined, the Far East market continues to grow. With a new model almost every week, makers like Suzuki, Honda, and Yamaha satisfy their customers' appetites with these plastic putt-putts. This Suzuki model followed in the footsteps of the earlier Address model which was available in Japan and Europe in the early 1990s.

▸The Vespa is a *cosa*, but the Cosa isn't a Vespa
In its wisdom, Piaggio discontinued the Vespa in 1985 and replaced it with the Cosa (meaning *Thing*), which was essentially the same scooter but with added Space-Age design. So much uproar ensued that Piaggio was forced to bring back the Vespa to cheering crowds. The Cosa came in three flavors: 125cc, 150cc, and 200cc all riding on 4.00x10in tires with 12-volt electrics. The 200cc scooter could exceed 65mph. The major change between the Vespa and the Cosa is the emblem on the legshield, while most parts are still interchangeable.

"There is a certain residual lawn mower likeness to the exhaust note, I admit, but otherwise the Helix is more motorcycle than scooter."
—David Edwards,
Cycle 1995

▸The Scooter of the Future?
BMW didn't believe the hype of the 1950s scooter craze, and although it built a prototype scooter in 1954, it didn't throw its hat into the ring and begin production. Then in 1992, BMW showed the C1 scooter at the Cologne Motorcycle Show, which it declared to be a new kind of vehicle not a car, not a scooter, but somewhere between the two. The C1 boasted all the safety of a car with a roll bar and four-point seatbelt but maintained the easy parking and maneuverability of a scooter. Two different engines were featured at the show, a 125cc and a hefty 250cc.

Third World Motorscooters

Mass Mobilization All Over Again

With economic circumstances often similar to the U.S. during the Great Depression and postwar Europe, countries from India to Taiwan needed cheap transportation, and European manufacturers unloaded their antiquated scooter tools and dies.

In 1972, Scooters India bought Innocenti's scooter tooling and has continued manufacturing late-1960 Lambretta models, all the while increasing horsepower, speed, and efficiency. Following their lead, Zündapp sold off their stock and tools in 1985 to China.

Not to be outdone by rival Lambrettas and other makes, PGO of Taiwan makes scooters which resemble Vespas, and Bajaj of India now pumps out more Vespas than Piaggio in Italy.

"Taiwan is the motorscooter capital of the world."
—*Popular Science*, May 1994

▶ **Argentine Repair Manual**
One of the main reasons that the Third World has remained scooter country is that the mechanical beasts are never laid to rest. They are continually repaired through any means possible, and if they are totaled, every part can be reused to revive another aching scooter. This manual was printed in 1965—ten years after the debut of the Lambretta TV 175—but still failed to put a true TV on the cover, opting instead for an LD.

▲ Ape in Sri Lanka
Three-wheeled Piaggio Apes (or "Bee") became the functional Vespa. Horses were replaced by Vespas for their efficiency and easy maintenance, but for transporting goods or taxi passengers the Ape took over where the rickshaw left off. This green Piaggio navigated the dangerous streets of Colombo, Sri Lanka, with no regard for "No Parking" signs.

▲ Siambrettas and Iso Divas Galore!
Pedestrians walk by these scooters as Americans walk by another Ford Taurus, uninterested. In the foreground a Siambretta LD missing its spare tire and chrome sidepanel piece is accompanied by a pair of Iso Divas and a string of other mopeds, motorcycles and other scooter wannabes.

▲ Proof of East Bloc Collusion!
Jawa and CZ were merged under state control in 1945. The Communist board ordered them to make a scooter, which appeared in 1946 as the Czeta, the pug-nosed wonder. *Sasha Wellborn*

◄ Classic Remnants
Two classic Italian scooters which are now seldom seen out on the streets of Rome are parked unceremoniously on a sidewalk in Salta, Argentina. The first is, a chromed Lambretta D sans paneling which would have been the LD, or Lusso, model; the second, a Rumi Formichino, or "little ant" with a two-cylinder 125cc engine that could leave many 250cc motorcycles of its era in the dust to breathe its two-stroke exhaust.

Snake Oil Scooter Repairs

It Came From Within the Garage....

The sound and the fury—or is that the sound and the worry? Sad but true, motorscooters do break down despite the promise of the maker that the miles would continue to unroll beneath your dwarf wheels—trouble-free mile after trouble-free mile, forever and ever, until death do you part.

When a scooter broke, it was often "repaired" by a ham-handed mechanic—usually its owner—who had never had the side cover off until that sad, fateful day. Naturally, quick-and-easy remedies to keep their beloved machines on the road were developed by scooterists around the world. Here's some of the best of the worst—with a caveat to perform these miracle cures at your own risk as none of them are recommended by these authors who have nary a rounded bolthead on their Lambrettas.

▲ Fun and Carefree Days Ahead
Repair Tip #1: When those fun and carefree days on your Vespa suddenly come to an end as you fly over the handlebars due to a seized engine—which is usually just a matter of time with a two-stroke scooter—let the overcooked engine cool down for a good fifteen minutes. Then put the scooter into second gear, tie a clothesline to your friend's Plymouth Volare, and tow the scooter behind the car until the engine frees up. It usually only takes a block or two. If this age-old cure does not work, let the scooter cool its heels overnight and try again. There are two other routes to repair seized engines: 1. Rebuild the thing, or 2. Make an offer on the Volare.

▲ Waiting For Help on the Side of the Road
Repair Tips #2 and #3: Always dress your best for success when riding off over hill and dale on your scooter. This way, you won't scare off anyone who may stop to help you when you find yourself on the side of the road. On the other hand, when the clutch in your Lambretta 75 finally sands itself smooth after a couple weeks of use, the repair is as simple as a can of Coca-Cola. Merely dump the contents onto the clutch and let sit for fifteen minutes. The super-sugarized soda pop will have your clutch working again as soon as it dries. This will not work for more permanent repairs, of course. To truly fix a clutch, you must let the Coca-Cola dry overnight.

► If All Else Fails, Paint It
Repair Tip #6: One of the best tuning tricks for scooters is paint. If your brakes wheeze with asthma, your engine farts black smoke, and rust blossoms like cancer on the rest of the scooter, simply buy a can of spray paint and shoot. The results will wow your friends, making them ask, When did you get the new scoot? The trick's on them of course, just as many people would swear this Agrati Capri scooter on the island of Naxos, Greece, is brand spankin' new—especially with its tuned exhaust.

▸ **Hey, Amico, Wanna Buy a Scooter?**
Repair Tip #5: Would you buy a used scooter from this man? "This Vespa is the one so-and-so raced across the Sahara Desert breaking the world record in 1955, but for you, my friend, a special price reflecting its special value." Always think before buying that special scooter—even from your special friend.

▴ **And If New Paint Fails, Take The Train**
Repair Tip #7: It would be just your luck, of course, to have your freshly painted Lambretta's engine eat itself for lunch outside this train station in Lambrate, the Milan suburb that was once home and namesake to the Innocenti factory that created your marvelous scooter. Quietly park your sizzling putt-putt behind a bush, buy a ticket home, and hope no Vespisti are watching.

▴ **Full-Race Lambretta LC 125**
Repair Tip #4: There's nothing like the tinny roar of a two-stroke engine about to detonate after a loud downshift in front of the local coffeeshop watering hole. But to keep your scooter riding that razor's edge of high performance you must now and then perform what's called a tune-up. A "tune-up" in motorscooter vernacular is needed when the scooter isn't running loud enough. The traditional cure is simple, effective, and cheap. Get a saw and cut off a good 6in of the exhaust pipe. This trick two-stroke tuning tip has been perfected over the decades by scooteristi everywhere, making for a scooter that not only sounds louder but is, of course, faster. The other route to full-race glory is to bolt on a motorcycle megaphone exhaust, as shown on this Lambretta LC 125 spotted in Salta, Argentina. Megaphones create a blunderbuss of orgasmic aural ecstasy for the rider and endear scooters to passersby. Get one today!

The Scooters that Never Were

The World Could Have Been a Better Place....

It's true, the world could have been a better place if these scooters had seen the light of day. Alas, they may not have brought peace, love, and understanding, but at least the world would have been a more interesting place for their mere existence.

"You can tell the happy scooterists by the bugs on their teeth."
—Famous proverb

◄ 1935 Gyroscope-on-Wheels
Los Angeles inventor Walter Nilsson created his Uniscooter in 1935 that was part motorcycle (due to its large wheels), part scooter (due to its small wheels), and part motorized gyroscope (due to everything else). Nilsson's one-wheeler actually operated like a wheel within a wheel—except that it incorporated four wheels in all. The outer wheel was driven by the one-cylinder engine, which propelled the wheel by gear cogs; the three small wheels worked like runners. The driver sat upright inside the contraption and steered with the auto-like steering wheel. According to notes on the flipside of this newspaper photo, the Uniscooter was steered "by a secret device that causes the wheel to lean and thereby turn while allowing the rider to keep upright." Nilsson drove his machine to a top speed of 18mph in second gear; he reported that higher speeds were feasible if he only had a custom pneumatic tire, which would set him back $800. Total cost of this invention was $5,000—in 1935! Buck Rogers must have been green with envy. *Herb Singe Collection*

◄ **1940s Lawn Mower Scooter**
No, this is not a rare still from the lost Shirley Temple movie, *Saving the Farm*. Rarer still, this is the Missing Link in the development of the riding lawn mower, a mower attachment for your motorscooter that "even a child can use," according to this press-release photo. Somehow, some way, you simply removed the front end of your scooter and bolted on this mower, turning your scooter into a four-wheeled goat, a precursor to today's riding mowers. Only in America. *Herb Singe Collection*

▲ **NSU Double-Scooter Car**
NSU was always a creative company. In 1952, the far-seeing NSU inventors took out their welding torches and fabricated this three-wheeled NSU-Lambretta designed for "those persons who like to keep three wheels on the ground." In 1953, those NSU visionaries were back with their recharged welding torches to cobble up a prototype car made from two NSU-Lambretta 125 scooters welded together side by side. Bathtub-like rear bodywork was added on to allow for four saddle seats. The handlebars on the left-side scooter had a bar running across to the steering rod of the right-side scooter so the two front wheels could be controlled by the one set of handlebars. This really happened. *Herb Singe Archives*

▲ **Shriner Super Scooter**
Leave it to those men in the red fezzes to dream up this eighteen-Shriner super scooter to dazzle the crowds at Fourth of July parades. This custom-built long-load putt-putt is the pride of the Long Riders from the Tripoli Temple. *Courtesy Noble Tom Rousseau, Imperial Photographer*

Scooter Sources & Suggested Reading

American Scooters
Antique Motorcycle Club of America
Dick Winger
PO Box 333
Sweeter, IN 46987 USA

Vintage Motor Bike Club
Joyce Lee, Secretary-Treasurer
537 West Huntington
Montpelier, IN 47359 USA
US club for "out-of-production" motor bikes and scooters, from Cushman to Whizzer, Simplex to Salsbury

Classic Scooters of America
Steven Zasueta & Edwin Moore, Editors
PO Box 152366
San Diego, CA 92195 USA

American Scooter Road Racing Association
6244 University Avenue
San Diego, CA 92115 USA

Topper Motorscooter Owners Club
456 W. 14th Street
Chicago Heights, IL 60411 USA
708-481-9685

Benelli
Benelli Owners Club
Dennis McGuire
3215 West 25th Street
Erie, PA 16506 USA

British Scooters
British Two Stroke Club
76 Hatch Road
Pilgrims Hatch
Brentwood, Essex, England
Devoted to all British two-stroke machinery

Cushman
OMC/Cushman, Inc.
PO Box 82409
Lincoln, NE 68501 USA

Cushman Club of America
Tom O'Hara, secretary
PO Box 661
Union Springs, AL 36089 USA

Ray Gabbard
Route 6, Box 95
Portland, IN 47371 USA
Cushman information and parts

Cushman Publications
Bill Somerville
1200 Kygar Road
Ponca City, OK 74604 USA
Cushman history and scrapbooks, shop and service manuals

Cushman Reproductions
Dennis Carpenter
PO Box 26398
Charlotte, NC 28221 USA
Cushman parts, accessories, tires, manuals, and more

Roger McLaren
193 3rd Street
Excelsior, MN 55331 USA
Cushman restorations and parts

Rich Suski
7061 County Road 108
Town Creek, AL 35672 USA
Cushman parts, accessories, tires, manuals and more, as well as parts for other vintage American motorbikes and scooters

Sam Kerley
Cushman Sales
6375 11th Street
Rockford, IL 61109 USA
Cushman parts

Coker Tire
1317 Chestnut Street
Chattanooga, TN 37402 USA
Blackwall and whitewall tires for Cushmans

Fox Grip
5181 Greencroft Drive
Dayton, OH 45426 USA
Reproduction handgrips, decals, and more for classic motorscooters

Paul Covert
1106 Alpine Lane
Dotham, AL 36301 USA
Cushman scooter parts as well as Truckster and Golfster parts

Anderson Manufacturing
7902 Shady Arbor
Houston, TX 77040 USA
White Port-a-walls for Cushman tires

Doodle Bug
Jim's Scooter Parts
Jim Kilau
593 North Snelling Avenue
St. Paul, MN 55104 USA
Doodle Bug parts and manuals, as well as Cushman parts and manuals

Ducati
International Ducati Owners Club
6061 Collins Avenue, Suite 16-C
Miami Beach, FL 33140 USA
American Ducati club devoted primarily to the motorcycles

Fuji
Erik's Scooter Service
3235 Byron Street
San Diego, CA 92106 USA
Rabbit parts and service

Guzzi
Moto Guzzi National Owner's Club
PO Box 98
Olmitz, KS 67564 USA
Excellent active club with good, technical newsletter

Harley-Davidson
Harley-Davidson Inc.
3700 West Juneau Avenue
Milwaukee, WI 53201 USA

Heinkel
Deutsches Motorräd Register
W. Conway Link
8663 Grover Place
Shreveport, LA 71115 USA
The vintage German motorcycle owner's association with excellent quarterly newsletter full of history and classified ads

Blue Moon Cycle
5711 Wood Valley Trail
Norcross, GA 30071 USA
Heinkel and other German scooter parts

Innocenti
Cosmopolitan Motors
301 Jacksonville Road
Hatboro, PA 19040 USA
US importer of Lambretta and MV Agusta scooters for a short period; small selection of parts available

West Coast Lambretta Works
Vince Mross
6244 University Avenue
San Diego, CA 92115 USA
Performance accessories, racing, and some restoration parts for Lambrettas

Vittorio Tessera
Via Marconi, 8
20090 Rodano (MI) Italy
Premiere source for information on vintage Lambrettas and Vespas, as well as most other types of European scooters. Also operates a Lambretta museum and runs the Italian Lambretta Club and Registro Storico

Stefano Panciroli
Via Casali, 8
42100 Reggio Emilia, Italy
Probably the world's best source of vintage decals for scooters and motorcycles of all makes, although specializing in European manufacturers

Lambretta Club and Registro Storico
Casella Postale n. 1
20067 Paullo (MI) Italy
Operated by Vittorio Tessera in conjunction with his Lambretta museum

Italian Scooters
Stefano Francalanci
Borgo S. Croce, 17
50122 Firenze, Italy
All Italian scooters and *ciclomotori*

Japanese Scooters
Vintage Japanese Motorcycle Club
John Sullivan, Jr.
117 Summit Drive
Bellefontaine, OH 43311 USA

Maico
Maico Owners Club
c/o Paul Hingston
"No Elms," Goosey, Nr. Faringdon
Oxon SN7 8OA England

Mondial
Moto Club Milano/
Mondial International Club
Via G. Washington, 33
Milan, Italy
FB Mondial club run with support from the factory's founding Boselli family

Montesa
Spanish Motorcycle Owners Group
168 Covington Court
San Jose, CA 95136 USA
All Spanish motorcycles and scooters

Mustang
Mustang Motorcycle Club of America
Alan Wenzel, Vice President
530 South Industrial Boulevard
Dallas, TX 75207 USA
US club for Mustangs

Mustang Motorcycle Registry
Steve Counter
3720 Flood Street
Simi, CA 93063 USA

MV Agusta
MV Agusta Club of America
Box 185
Wiscasset, ME 04578 USA

NSU
NSU Club of America
717 North 68th Street
Seattle, WA 98103 USA

Parilla
Parilla International Registry/
Registro Storico Moto Parilla
3620 South 35th Avenue
Minneapolis, MN 55406 USA
Owners club for all Parilla scooters and motorcycles with newsletter covering Parilla history and restoration

Piaggio
Vespa, U.S.A.
1709 North Orangethorpe Parkway
Anaheim, CA 92801 USA
The oldest, most established shop in the USA for Vespa sales, parts, and repair services

Vespa Supershop Inc
2525 University Avenue
San Diego, CA 92104 USA
Vespa parts as well as Malossi carburetors and other high-performance equipment

First Kick Scooters
1318 4th Street
Berkeley, CA 94710 USA
Specializes in vintage and modern Vespas with mail-order parts service

Vespa Motorsport
3450 Adams Avenue
San Diego, CA 92116 USA
Vespa parts and service

Scooterworks USA
7117 North Clark Street
Chicago, IL 60626 USA
Original parts for all Vespas as well as restoration service and sales of vintage Vespas

Vespa of Chicago
6609 North Clark Street
Chicago, IL 60626 USA
US importer of Vespas for twenty years. Large stock of new scooters and parts supply

Scooterville USA
1709 North Orangethorpe
Anaheim, CA 92801 USA
Vespa parts
Vespa Club of America
PO Box 670864
Dallas, TX 75367 USA

Sears Allstate Owners Club
3060 Stoneycreek
North Royalton, OH 44133 USA

Federation of International Vespa Clubs/USA
Rolf Soltau
1566 Capri Drive
Campbell, CA 95008 USA

Powell
Powell Cycle Registry
c/o Wallace Skyrman
4588 Pacific Highway
North Central Point, OR 97502 USA
Powell history, information, and parts literature

Rumi
Registro Storico Italiano Moto Rumi
Riccardo Crippa
Via Palma il Vecchio, 3
24100 Bergamo, Italy
Premiere source for Moto Rumi arts and the Italian Moto Rumi club

lsbury
rb Singe
0 Central Avenue
llside, NJ 07205 USA

rrot
rrot Club—Section Motos
rue Lamartine
100 Thouars, France

iumph/BSA
umph International Owners Club
Box 6676
lliston, MA 01746 USA

ündapp
ündapp Bella Club
/o B. Sauer/K. Reiken
AN Muhlgraben 14
D-5439 Bad Marienberg Langenbach
Germany

Scooter Memorabilia
Scootermania!
235 Byron Street
San Diego, CA 92106 USA
Catalog of vintage scooter photographs suitable for framing—a must-have for all self-respecting Vespisti and Lambrettisti.

Scooter Magazines and Books
MotorScooter magazine
PO Box 2012
Rancho Cucamonga, CA 91729 USA

Scooter Magazine
Editrice L'Isola
piazza Roma, 1
22070 Lurago Marinone (CO) Italy

Hemmings Motor News
PO Box 1108
Bennington, VT 05201 USA

Classic Motorbooks/
Motorbooks International
PO Box 1
Osceola, WI 54020 USA
1-800-826-6600

Suggested Reading

Barnes, Richard.
Mods!
London: Eel Pie Publishing, 1979.
Vastly entertaining photographic history of the Mod movement and the Mod-Rocker wars.

Brockway, Eric.
Vespa: An Illustrated History.
Sparkford, England:
Haynes Publications, 1993.
Should read *Douglas* Vespa: An Illustrated History *of English Vespas.* Even with the correct title there's not much to recommend it, I'm afraid.

Brown, Gareth.
Scooter Boys.
English rebel scooter culture in lots o' candid B&W photos by the editor of England's *Scootering* mag.

Crippa, Riccardo.
Rumi: La moto dell'artista.
Milan, Italy: Giorgio Nada Editore, 1992.
Probably the only book that will ever need to be written about Rumi.

Dregni, Michael and Eric Dregni.
Illustrated MotorScooter Buyer's Guide.
Osceola, WI:
Motorbooks International, 1993.
Overly analytical, pseudo scholarly, pompously turgid, and undoubtedly libelous. Well, *we* liked it.

Dumas, François-Marie and Didier Ganneau.
Scooters du monde.
In press.
Color and B&W photos of scooters around the world. Sure to be good.

Gerald, Michael.
Mustang: A Different Breed of Steed.
N.p.: self-published, n.d.
The definitive work on Mustang.

Goyard, Jean and Dominique Pascal.
Tous les scooter du monde.
Paris: Editions Ch. Massin, 1988.
The *du monde* part of the title is misleading as American scooters are hardly mentioned, but this is the best encyclopedia of European scooters and scooter culture available.

Goyard, Jean, Dominique Pascal, and Bernard Salvat.
Vespa Histoire et Technique.
Paris: Editions Moto Legende/Rétroviseur, 1992.
It's big, it's beautiful, it's all about Vespas.

Hebdige, Dick.
Hiding in the Light.
London and New York: Routledge, 1988.
Includes a fun, insightful, yet turgid chapter on "Object as Image: the Italian Scooter Cycle."

Herlingshaw, Ken.
The Lambretta Manual of Performance Tuning and Conversions.
Croydon, England: Kingsley Press, n.d.
Excellent in-depth guide to high-performance modifications for mid-1960s to 1971 Lambrettas.

Kubisch, Ulrich, editor.
Deutsche Motorroller 1949-73.
München, Germany:
Schrader Automobil-Bücher, 1992.
Collection of German scooter ads. Auf deutsch.

Kubisch, Ulrich.
Vespa mi' amore.
Hösseringen, Germany:
Schrader verlag, 1993.
Color and B&W pictures that never end of Vespas around the world, as well as pioneering scooters and most other makes.

Lambretta Story. (Video cassette).
Milan, Italy: Giorgio Nada Editore, 1994.
Fun yet brief documentary history of the Lambretta in films, culture, and collections.

Lintelmann, Reinhard.
Deutsche Roller und Kleinwagen der Fünfziger Jahre.
Brilon, Germany:
Podszun Motor-Bücher, 1986.
Without doubt, *the* history of German scooters, micro cars, and other fascinating Teutonic oddities—auf deutsch.

Pascal, Dominique.
Scooters de chez nous.
Boulogne, France: Editions MDM, 1993.
Fun historical photo book of scooters in France.

Rivola, Luigi.
Chi Vespa mangia le mele: Storia della Vespa.
Milan: Giorgio Nada Editore, 1993.
A blend of potted Piaggio history and how-to workshop manual with specifications of all Vespa models. In italiano.

Roos, Peter.
Vespa bella donna.
Kiel, Germany: Nieswand Verlag, 1990.
Reprints Vespa calendar girl pictures in full color.

Roos, Peter.
Vespa Stracciatella: Ein Lust- und Bilderbuch von der italienischen Beweglichkeit.
Berlin: Transit Buchverlag, 1985.
Germanic Vespa culture with tiny photos and poor reproduction—but a nice cover! Auf deutsch.

Sartre, Jean-Paul.
Being and Nothingness.
1943.

Somerville, Bill.
A History of the Cushman Eagle.
Ponca City, OK: Cushman Publications, n.d.
The history of the Eagle. Good stuff.

Somerville, Bill.
A History of the Cushman Motor Works.
Ponca City, OK: Cushman Publications, 1986.
A fine overview of Cushman lore.

Somerville, Bill.
Allstate Scooters & Cycles 1951-1961.
Ponca City, Oklahoma: Cushman Publications, 1990.

Somerville, Bill.
The Complete Guide to Cushman Motor Scooters.
Ponca City, Oklahoma: Cushman Publications, 1988.
The history of Cushman, lots of rare photos, and extensive tables listing engine and models. Recommended.

Stuart, Johnny.
Rockers!
London: Plexus Publishing Ltd., 1987.
This is a great book, the counterpart to Richard Barnes' *Mods!* Don't be a snob, get it.

Tessera, Vittorio.
History of Lambretta.
Milan: Giorgio Nada Editore, in press.
He's Senor Lambretta and a swell guy, too. It's going to be great.

Tessera, Vittorio.
Scooters Made in Italy.
Milan: Giorgio Nada Editore, 1993.
An incredible encyclopedia of Italian scooters from rare prototypes through the Vespa and Lambretta lines. In italiano.

Tessera, Vittorio.
Lambretta 1947-72.
N.p.: self-published, n.d.
Photocopied Lambretta model history with specifications.

Tessera, Vittorio.
Vespa 1946-72.
N.p.: self-published, n.d.
Photocopied Vespa model history with specifications.

Vespa Story. (Video cassette).
Milan, Italy: Giorgio Nada Editore, 1992.
Fun yet brief documentary history of the Vespa in films, culture, and collections.

Vespa Style Handbook.
N.p.: Fukimi Publishing, 1992.
Lots o' color and pics of Japanese Mods. Good, clean fun.

Walker, Mick.
Italian Motorcycles.
Huddersfield, England: Aston Publications, 1991.
The best overview of Italian cycles and scooters available. Lots o' pictures.

Webster, Michael.
Motorscooters.
Haverfordwest, England: Shire Publications, Ltd., 1986.
Excellent overview with lots of photos.

Zeichner, Walter, ed.
Vespa Motorroller 1948-1986.
München, Germany: Schrader Automobil-Bücher, 1987.
Reprints of Hoffmann and Messerschmitt Vespa ads.

Index